1994

University of St. Francis
GEN   581.632 D874
Duke, James A.,

T3-BNH-595

00076453 6

Handbook of
Edible Weeds

Handbook of
Edible Weeds

Author

James A. Duke

College of St. Francis Library
Joliet, Illinois

CRC Press

Boca Raton   Ann Arbor   London   Tokyo

Library of Congress Cataloging-in-Publication Data

Duke, James A., 1929-
  Handbook of edible weeds / author, James A. Duke.
    p.   cm.
  Includes bibliographical references and index.
  ISBN 0-8493-4225-2
  1. Wild plants, Edible--United States--Handbooks. manuals, etc.
  2. Weeds--United States--Handbooks, manuals, etc.   I. Title.
QK98.5.U6D85   1992
581.6'32--dc20   91-31934
                                                                        CIP

This book represents information obtained from authentic and highly regarded sources. Re-printed material is quoted with permission, and sources are indicated. A wide variety of references are listed. Every reasonable effort has been made to give reliable data and information, but the authors and the publisher cannot assume responsibility for the validity of all materials or for the consequences of their use.

All rights reserved. This book, or any parts thereof, may not be reproduced in any form without written consent from the publisher.

Direct all inquiries to CRC Press, Inc., 2000 Corporate Blvd., N.W., Boca Raton, Florida 33431.

© 1992 by CRC Press, Inc.

International Standard Book Number 0-8493-4225-2

Library of Congress Card Number 91-31934

Printed in the United States of America   2 3 4 5 6 7 8 9 0

581.632
D874

PREFACE

More than five years in the making, this book culminates my long-standing interest in weeds, economic plants, and foraging, here combining all of these. It should be added that, although I have eaten at least one of the species in each of the 100 genera covered and suffered no ill consequences from them, except as noted for daylily and violet, I cannot guarantee that the next person might not be seriously allergic to them. Just before writing this Preface, I talked with one scientist whose wife is violently, potentially fatally, allergic to buckwheat. Buckwheat is related to the sorrels and docks. Conceivably she might be even more allergic to other members of the buckwheat family, like dock, Mexican bamboo, rhubarb, sorrels, tearthumbs, and wild buckwheats, some of which are discussed in this book. Because of the possibility of allergic reactions, readers are warned of this possibility, and reminded that neither my employers, sponsors, publishers, nor I can endorse grazing or foraging, even when under the tutelage of experienced botanists. *SOMEBODY, SOMEWHERE, MAY BE FATALLY ALLERGIC TO EVEN OUR TAME SPECIES, LIKE PEANUTS AND PECANS. THEY MIGHT BE EQUALLY ALLERGIC TO WILD RELATIVES OF THESE OR OTHER SPECIES.* Caution is advised in sampling any new food, if you are going to be brazen enough to indulge. If you still insist on proceeding, you should first be sure of the identity of the plant and then very cautiously sample, testing your reactions thereto. Experiment at your own risk!

Having frightened you half to death with the caveat, let me remind you that the Surgeon General, while not recommending these weeds specifically, has recommended that we increase the relative quantity and variety of fruits, grains, and vegetables in our diets. Many edible weeds are rich, not only in vitamins and minerals, but also in more esoteric compounds that the National Cancer Institute (NCI) is now studying for their potential as cancer preventives. Though the NCI has focused on several compounds that occur in soybean, these same compounds occur also in many of our wild legumes, for example, *Apios* and *Amphicarpaea,* discussed in this book.

North American Indians had at least 1112 food plant species (among them *Apios* and *Amphicarpaea*), more than most of our readers will encounter in a lifetime, travelling all the continents and sampling all the cuisines. A modern supermarket may have as many as 600 varieties of greens, vegetables, fruits, grains and nuts. Still they will miss most of the weeds mentioned in this book, though a few have found their way to a few supermarket shelves, like species of *Amaranthus, Chenopodium,* and *Nasturtium.* Remember, moderation in all things, except, perhaps, dietary diversity! Vary your diet judiciously. Don't follow the advice of those who say eat only fruits, or eat only raw vegetables, or eat only locally grown organic whole grains. Growing your own and "grazing", as some people call foraging for wild foods, can be fun if you know what you're doing. Enjoy. But carefully!

157,426

THE AUTHOR

James A. "Jim" Duke is a Phi Beta Kappa Ph.D. (botany, University of North Carolina, 1961). Following military service, Jim undertook postdoctoral curatorial and teaching activities at Washington University and Missouri Botanical Garden in St. Louis, Missouri. There he began studies of neotropical ethnobotany, his overriding interest to this day. From 1963 to 1965, Duke was ecologist for the USDA (Beltsville, Maryland), joining Battelle Columbus Laboratories (1965 to 1971) for ecological and ethnobotanical studies in Panama and Colombia. Rejoining USDA in 1971, Duke had assignments relating to crop diversification, and economic-, energy- and medicinal-plant studies in developing countries.

Fluent in Spanish and dabbling with Portuguese, Duke has studied and/or lectured widely, concentrating on tropical ecology, medical botany, and crop diversification. Duke "cut his tropical eye teeth" in Panama, where he was resident from 1966 to 1968. Currently, he is an Economic Botanist with the National Germplasm Resources Laboratory, Agricultural Research Service, USDA, Beltsville, Maryland. He is preparing an encyclopedia of economic plants and collaborating with the National Cancer Institute on both their AIDS and cancer-screening programs and their new Designer Food Program (to prevent cancer). His databases on the ecology, nutritional content, folk medicinal uses, and chemical constituents of 1000 economic plants are being widely utilized. CRC will be publishing his phytochemical databases shortly. Duke's major goal lately is to reverse the disdain for alternative medicines in the U.S. where, as in the Third World, a larger and larger percentage of the people can no longer afford First-World pharmaceuticals. This field guide underscores Duke's contagious interest in natural foods and nutritional approaches to preventive medicine.

Dr. Duke belongs to the American Botanical Council (Trustee), American Herb Association (Life), American Society of Pharmacognosy, Associates of the National Agricultural Library, Association for Tropical Biology (Life), Council of Agricultural Science and Technology (Cornerstone Life Member), Herb Reseach Foundation (Advisor), International Association of Plant Taxonomists (Life), International Society for Tropical Root Crops (Life), International Weed Science Society (Life), Organization for Tropical Studies (Life), Oriental Healing Arts Society (Honorary), Programa Interciencia de Recursos Biologicos (Advisory Council), Sigma Xi, Smithsonian Institution (Collaborator), Society for Conservation Biology (Life), Society for Economic Botany (Life), Southern Appalachian Botanical Club (Life), Tri-State Bluegrass Association (Life), and the Washington Academy of Science (Life).

In addition to scores of popular and scientific articles, Duke has published several pertinent books: (1) *Handbook of Legumes of World Economic Importance* (Plenum Press); (2) *Medicinal Plants of the Bible* (Trado-Medic Books); (3) *CRC Handbook of Medicinal Herbs* (CRC Press); (4) *Culinary*

Herbs: A Potpourri (Trado-Medic Books); (5) *Medicinal Plants of China* (with E. Ayensu Reference Publications); (6) *CRC Handbook of Proximate Analysis Tables of Higher Plants* (with A. Atchley; CRC Press); (7) *Isthmian Ethnobotanical Dictionary* (3rd edition, Scientific Publishers, Jodhpur, India); (8) *Handbook of Northeastern Indian Medicinal Plants (Quarterman Press); (9) Living Liqueurs* (Quarterman Press); (10) *CRC Handbook of Agricultural Energy Potential for Developing Countries* (with A. Atchley, K. Ackerson, and P. Duke; CRC Press); (11) *CRC Handbook of Nuts* (CRC Press); (12) with Steven Foster, a Peterson *Field Guide to Medicinal Plants* (Houghton-Miflin); and (13) *Ginseng, a Concise Handbook* (Reference Publications). Duke is a rather regular contributor or contributing editor to such periodicals as *American Health, Business of Herbs, Coltsfoot, Diversity, Economic Botany, The Environmentarian, HerbalGram, Organic Gardening, The International Permaculture Species Yearbook,* and *Lloydia (Journal of Natural Products).*

A popular lecturer on ethnobotany, herbs, medicinal plants, and their ecology, Dr. Duke has taped dozens of TV and radio shows (and a video history of his career is in the National Agricultural Library!). With his wife and illustrator, Peggy, he grows a number of interesting plants on his six-acre farmette (Herbal Vineyard).

ACKNOWLEDGMENTS AND DEDICATIONS TO THE ILLUSTRATORS

Having known, for more than two decades, two of the artists whose line illustrations contribute so much to this text, I acknowledge with great respect the importance of art in conveying a message, in this case, mental from visual pictures of weeds. Peggy Kessler Duke, M.A. Botany, University of North Carolina, Chapel Hill (1956), brings to the book about half the illustrations. After taking her M.A. degree, Peggy moved to St. Louis, where she married yours truly, Jim Duke, in 1960.

For this book, Peggy's drawings are mostly original, but some were redrawn, with permission, from other books she illustrated, e.g., Hollis G. Bedell's *Vascular Plant Taxonomy Laboratory Manual* (Department of Botany, University of Maryland, College Park, 1985, 174 pp.) and Melvin L. Brown and Russell G. Brown's *Herbaceous Plants of Maryland* (Port City Press, Baltimore, 1984, 1127 pp.), and my *CRC Handbook of Nuts* (CRC Press, Boca Raton, FL, 1989, 343 pp.). Peggy also consulted some of the marvelous unpublished photographs of photographer Neil Soderstrom and illustrations in several old public domain books in preparing some of these drawings. Like the author, the artist is often surrounded by source books in preparing a picture, combining the best of many sources, including the living plants, many of which grow spontaneously or under cultivation at Herbal Vineyard.

Some of Peggy's public domain drawings first appeared in Steven R. Hill and Peggy K. Duke's *100 Poisonous Plants of Maryland* (Bull. 314, Univ. Md. Coop. Extens. Serv., 1985—1986, 55 pp.). The public-domain drawing of *Arundo* was taken from A. S. Hitchcock and A. Chase's *Manual of The Grasses of the United States* (USDA Misc. Publ. No. 200, 1950, 1051 pp.). Drawings of *Acorus, Asclepias, Fragaria, Gaultheria,* and *Lycopus* were taken from *A Guide to Medicinal Plants of Appalachia,* by A. Krochmal, R. S. Walters, and R. M. Doughty (USDA Agric. Handbook No. 400, 1969, 291 pp.). According to the acknowledgments in the last-named volume, "Mrs. Marion Sheehan, Gainesville, Florida, created most of the line drawings used" in Handbook 400, A Guide to Medicinal Plants of Appalachia.

Several illustrations by the great Regina O. Hughes, acknowledged elsewhere, were taken from Clyde Reed's USDA Agric. Handbook No. 366, Selected Weeds of the United States. My friend Regina Hughes, since illustrating this USDA Handbook, as "*scientific illustrator,* Crops Research Division, Agricultural Research Service," has been awarded a well-deserved honorary doctorate. I am pleased, then, to dedicate this book, with gratitude, to these distinguished illustrators, alphabetically Mrs. Peggy Duke, Dr. Regina Hughes, and Mrs. Marion Sheehan. Hopefully your illustrious work will make many of us think twice before condemning our weeds. There is beauty among them, as your artwork shows.

TABLE OF CONTENTS

Introduction

INTRODUCTION

If, instead of spraying our weeds, we ate the safe ones, we would save all that energy tied up in the manufacture and application of pesticides (U.S. farmers spend an estimated $3 billion a year applying herbicides) and in the raising, processing, and shipping of more conventional foods. In this small book, I discuss the edibility of some of our more common weeds, many of which are quite nutritious and some of which have promising medicinal virtues as well. In their new Designer Food Program, the National Cancer Institute (NCI) is looking at some of these: burdock, garlic-mustard, watercress and other cresses, and wild garlic and other onion relatives, as cancer preventives or sources of cancer-preventive compounds. It seems appropriate in this first year of the final decade of the 20th century that the NCI is finally involved in looking for preventives for cancer. The return for taxpayer dollars should be greater there than in the search for the cure. The old wisdom still maintains! Eat plenty of fruits, vegetables, and grains, and eat a wider variety . . . and . . . . All things in moderation! Even weed-eating or grazing!

One common definition of a weed is a plant growing where it is not wanted. Thus the corn that volunteers in a soybean field is a weed, and vice versa. The dandelion, anathema to lovers of monocultural lawns, is much sought after by wine and potherb connoisseurs, not to mention goldfinches. Suburban gardeners and lawn cultists often battle many native American broadleaf species that invade their lawn, famous for its lack of biological diversity. Clearly more than half of Whitey Holm's "World's Worst 18 Weeds" (Holm et al., 1977) are edible, many treated here in this CRC *Handbook of Edible Weeds*. If you can't beat them, eat them!

> "Welcome to the Wonderful World of Weeds....With agricultural yield losses and control costs of $15.2 billion annually (in the U.S. alone), weeds are certainly enough to give farmers — and gardeners — nightmares galore. About 1 of every 10 species of plants on the face of the earth is a weed — some 30,000 weed species in all. Eighteen hundred of those 30,000 cause serious economic losses in production of one crop or another, and about 300 weed species plague cultivated crops throughout the world. The United States has become home to 70 percent of the worst weeds in the world. U.S. farmers spend an estimated $2.1 billion annually for herbicides, plus another $938 million in application costs. Herbicides — chemicals designed specifically to fight weeds — account for more than 65 percent of all pesticide sales in the United States" (Hays, 1991).

Does the fact that costs of and losses to weeds go up every year suggest that we are winning or losing the battle against the evasive and evolving enemies? Weeds, like the malaria organism, tend to evolve quicker than our pesticides do. The malaria organism may go through a generation every 3 or 4 days, and a few weeds can go through a generation in 3 or 4 weeks. But it takes years to develop a new pesticide and get it approved.

One nonedible but slightly culinary herb not to be covered later in this book is *Artemisia annua,* "Sweet Annie" to us, "Qing Hao" to the Chinese. Quinine was Nature's first important contribution to the medicine chest against malaria. But the malaria organism soon developed a tolerance to quinine, and quinine no longer works in such cases. So we resorted to a major modification known as chloroquine. Now the malaria organism is developing a tolerance to chloroquine. So once again, we are knocking at Mother Nature's door, in search of a drug for chloroquine-resistant malaria. "Sweet Annie" may be Mother Nature's answer this time. I first met "Sweet Annie" on the Shenandoah River. My cousins were celebrating their 25th wedding anniversary, and my bluegrass group, then the "Howard County Dump", was participating. Les, the guitar player, was also a medical doctor working for Walter Reed Hospital, specializing in tropical diseases. The band played at the upper end of a canoe-livery dropoff point. Then we floated down the Shenandoah in several canoes with several six packs, the latter generating frequent trips to shore. On one "pitstop", I found myself surrounded by "Sweet Annie", then new to me, which I subsequently identified as *Artemisia annua*.

I forgot all about that nice party until a couple of years later when, in Kunming, China, a Chinese scientist asked me if I knew "Qing Hao". I told him about the dense stands of "Qing Hao" on the Shenandoah. He told me that "Qing Hao" was proving effective against chloroquine-resistant malaria. Since I was there to study anticancer plants, I was only marginally interested, but did add his comments to my field notes.

Another year or two passed and Les, the guitar player, called to ask if I knew *Artemisia annua*. My answer: "Les, do you remember that aromatic herb back at pit-stop number 2 at my cousins' anniversary party on the Shenandoah? That was *Artemisia annua*." The following weekend we went out and filled his car with "Sweet Annie" for a Walter Reed research program. "Sweet Annie" overpowered our mucous membranes on the way home and generated headaches. It is a powerful weed, like so many weeds filled with strong chemicals called secondary metabolites. These often enable the possessors to overcome bacterial, fungal, insect, and plant competitors. I no longer have to go to the Shenandoah to seek "Sweet Annie". She is a welcome weed at my place. With "Sweet Annie", the secondary metabolite is an endoperoxide called artemisinin, and it will kill the chloroquine-resistant malarial strain (today). If it passes clinical trials and becomes widely used for malaria as it already is in China, I predict that the malaria organism will develop a tolerance for it within 10 years. Then we will be back at Mother Nature's door, knocking again, seeking a drug for "Qing Hao"-resistant malaria. Let's hope that our ecocidal tendencies have not so reduced the biodiversity of our planet that we lost the species that held the answer to "Qing-Hao"-resistant malaria. Remember that 1 out of 10 species is a weed, 1 out of 10 species is endangered, at least 1 in 10 has edible parts (some estimate that there are 80,000 edible species among the 250,000 to 300,000 higher plant species), and 25% of known higher plants, weeds included, have published folk or proven medicinal applications (at least in China and the U.S.). However, fewer than 1 in 10 have antimalarial properties. The more endangered species we allow to perish, the less likely the antimalarial (or anti-AIDS or whatever) drug will be found in the remaining species. Another introduced weed, the stinktree or tree of heaven, *Ailanthus altissima,* also has at least five compounds which show *in vitro* antimalarial activity.

What is the value of *Artemisia annua* if its artemisinin saves the thousands of lives expected to be lost to malaria? Some have even suggested that the endoperoxide of "Sweet Annie", like that of the "epazote" *Chenopodium ambrosioides,* might make a special contribution to the deceleration of the greenhouse effect, providing oxygen in anerobic environments like landfills, rice paddies, ruminant tummies, and termite mounds, oxidizing malodorous methane to CO_2, ameliorating and sweetening, if not significantly decreasing, greenhouse gases. At concentrations less than 10 parts per million (ppm), artemisinin is one of Nature's herbicides as well. While I am not going to argue that natural herbicides are necessarily any safer than synthetic ones, I will argue that there is a better chance that my genes, over my evolutionary history, have been exposed to the natural pesticide than to tomorrow's synthetic pesticide. Weeds, like our higher plant foods, seem to have at least 10,000 times more natural pesticides than synthetic pesticide residues. Aromatic herbs and spices, like many weeds, may have 100,000 times more. For example, we can get 1,8-cineole from dozens of herbs and weeds, including "Qing Hao". 1,8-Cineole is a natural bactericide, herbicide, and insect repellant that occurs in dozens of our foods. I have elsewhere recommended that instead of spraying synthetic pesticides on our food plants, we remove natural pesticides as part of food processing, putting them into containers for use as pesticides. Our foods contain, naturally, many deleterious phytochemicals that we now synthesize from petroleum. As petroleum becomes more expensive, such phytochemicals may become economically competitive with petrochemicals, especially if the bulky biomass residues are converted into alternative fuels like butane, butanol, ethanol, methane, or methanol. Again, any time we use a living green plant as a fuel instead of using a fossil fuel, we decelerate the greenhouse effect. Another opportunistic photosynthetic plant moves into the spaces liberated as we harvest our energy plants, but such plants cannot survive without light in the coal mine or oil well.

The herbicide approach may seem a bit futile, remembering that Nature abhors a vacuum. If we eradicate one weed with an herbicide, there will be other aggressive species waiting to fill the void, some perhaps resistant to the herbicide. Many weeds are evolving resistances, not to just one, but to several herbicides. If I were seeking herbicide tolerances in weeds, I would look along long-sprayed railway rights-of-way. Those brown rights-of-way, like fields and fencerows following herbicides, violate the "Think green" approach to slow the "greenhouse effect". Any place we can replace asphalt, bare earth, barren desert, scorched earth, or sprayed vegetation with photo-synthesizing green plants, there is an increase in the world's oxygen production and a decrease in carbon dioxide production. The production of herbicides is energy intensive, replacing the energy of obese gardeners with the energy of the CO_2-generating factories and fossil fuels.

We import stramonium (the weed *Datura stramonium*) and belladonna from Europe as sources of scopolamine and atropine while we spray or otherwise wage battle with the jimsonweed, *Datura,* in our soybean fields. Monthly, we import more than a million dollars' worth of psyllium while waging war with our own *Plantago* weeds. We import another million dollars' worth monthly of senna, while waging biological warfare on our sicklepod senna. Our consumption of anticancer drugs has endangered the Asian mayapple species while ours languishes, all but unrecognized as a food and medicine source in our eastern deciduous forests. Our steroid contraceptives now derive mostly from soybeans, but could be started from weeds, even bamboos, with which we engage in chemical and biological warfare. There are dozens of phytochemicals which we import or synthesize that could be extracted renewably from weeds hand-pulled from our soybean and corn fields. We get most of our industrial chlorophyll from cultivated alfalfa or spinach. At the prices in the Sigma Chemical catalog for chlorophyll, each acre of weed contains millions of dollars worth of chlorophyll. An acre of jimsonweed could easily have 5 tons of biomass, of which 1 ton could be leaf. That ton of leaf could easily contain 0.1% or 1000 grams of chlorophyll. If that were extracted and purified (and as valuable as alfalfa or spinach chlorophyll), it would have a value of nearly $20 million dollars.

More a lover of the forest than of the lawn or the weeds with which it competes, I have a conflict of interest in writing this introductory harangue. For the last couple of years, I have been listing several economic products utilized here in the United States that could be obtained renewably from the rainforest. That same atropine, chlorophyll, and scopolamine mentioned above could be obtained from a rainforest *Brugmansia* or from our jimsonweed, named after Jamestown, Virginia. "Green" consumers might rather that atropine, chlorophyll, and scopolamine were extracted renewably from rainforest with profits reverting to the rainforest. "Buy American" consumers might rather they were derived from a North American weed species than from aliens like alfalfa, belladonna, qing hao, or spinach.

There are enough edible weeds (unfortunately often tainted with lead and other contaminants) in vacant lots and waste places in inner city America to correct many nutritional deficits. Even without eating the "weeds", there's enough food to feed the world (at least at 5 billion population), but it just can't be economically distributed to all starving people.

We use chemical energy to spray weeds, temporarily halting their photosynthetic conversion of carbon dioxide and water to sugars and oxygen. The dead plant decomposes, generating carbon dioxide and water. No more photosynthesis, hence no more CO_2-fixation, occurs until herbicide levels lower, through decomposition or lateral or vertical movement in the environment, to a level tolerated by the same or some other potential weed species. Herbicide manufacture generates CO_2, the decomposition of the dead weed generates CO_2, and the "scorched" earth fixes no CO_2 until an opportunistic plant, usually a weed, moves into the void Nature abhors. That's a threefold, admittedly minor, contribution to the greenhouse effect. If more than half of our worst weeds are edible, and instead of spraying them, we manually harvest and eat them, there will be no scorched earth. A new weed will move in shortly, weather permitting. The weed we eat substitutes for a more conventional food, grown in our backyard or grown as far off as California or Chile and shipped a great distance, with the generation of more CO_2. By substituting a locally grown weed for a vegetable shipped in, we add our fourth blow to the greenhouse effect. And we are heeding the Surgeon General's advice to eat less fat (especially fat meat) and to eat more fruits, vegetables, and grains, of wide variety.

Closing out 1990, I had three exciting finds about three edible weeds treated herein. *Science News* published an article suggesting that soy-lecithin could prevent cirrhosis, at least in gorillas. I consulted my father-nature's farmacy (FNF) database, and found that the despised dandelion flowers were reportedly higher in lecithin than any of the species for which I have quantitative data. Dandelion has a long folk history for hepatosis and jaundice, if not cirrhosis. Brazilian visitors, seeking high-cysteine legumes, were given roots of *Apios americana,* a legume which probably also contains lecithin (not to mention the chemopreventive Bowman-Burk inhibitors, phytic acid, isoflavones, saponins, and sterols, which the NCI is viewing as possible chemopreventives of cancer. Similar isoflavones occur in other edible weeds, for example, kudzu and red clover). My Brazilian friends, excited about the high cysteine content of *Apios,* then asked me to search the FNF database to see which plants were high in tryptophan, the mildly sedative amino acid that probably occurs in all higher plants. To my surprise, the richest source of tryptophan reported in FNF was in seeds of my favorite weed, the evening primrose, *Oenothera biennis.* Its seeds are a major source of the nutrient gamma-linolenic-acid (GLA); curiously the sprouting seeds have more alpha-linolenic-acid (ALA), found also in seeds of another weed, the *Perilla.* The Food and Drug Administration (FDA) has "busted" promoters of evening primrose at least three times in the last decade for making health claims that have not been substantiated to the satisfaction of the FDA. More recently, I am told, the FDA has banned tryptophan, following several deaths due to contaminated tryptophan. If all higher plants contain tryptophan, and I suspect they do, does that make all higher plants illegal? Thus, only in December of 1990 did I realize that dandelion was a rich source of lecithin, groundnut a rich source of cysteine, and evening primrose seeds a rich source of tryptophan. I have been consuming these weeds for a decade, with no apparent ill effects. The compounds cysteine, lecithin, and tryptophan are what I call ubiquicts, probably found in all higher plant species.

Table 1 was drawn from my upcoming FNF database (CRC Press), tabulating the phytochemicals reported from all GRAS (Generally Regarded as Safe) herbs and many food and medicinal plants, all told about 800 species. The abbreviation PL? means that the compound probably occurs in the plant; PL UBI means that the compound is probably ubiquitous in higher plants, the interrogation mark indicating speculation on my part. But it's an educated speculation. I'll wager heavily on those marked PL UBI and PL UBI? being found in the *Artemisia,* including the millions of dollars worth of chlorophyll. An acre of *Artemisia* could yield conservatively 5 tons dry biomass. Fertilization with appropriate sewage sludge could make unproductive waste places produce not 5 but 10 tons biomass per acre. All or many of the above chemicals and then some could be extracted from the biomass, still leaving behind a lot of biomass to be used in the production of ethanol, methane, or methanol, for energy purposes. This could be grown on flood plains. But it would take just as complex and eerie an array of retorts, distillers, and the like as one sees in the petrochemical complexes along our eastern turnpikes. I predict that local modest-scale phytochemical factories, fed with weeds and other plant residues, will be the energy sources of tomorrow, if there are no nuclear or solar breakthroughs.

This weed book commenced 5 years ago, when, with photographer Neil Soderstrom, I prepared a draft on edible plants of the East. I intentionally limited that draft to 100 eastern wild plants that I had personally consumed. Interestingly, 73% of those species were later defined as weeds in the revised Weed Science Society of America (WSSA) *COMPOSITE LIST OF WEEDS* (1989), a listing of 2076 weed species of current or potential importance in the U.S.

Alphabetically, here are the 27% of the edible genera (or species) not included in the WSSA list, some of which I have excluded from this book for that reason alone: *Amphicarpaea, Apios, Aplectrum, Aralia, Asarum, Caltha, Cardamine bulbosa, Carpinus, Castanea, Celtis, Cercis, Cryptotaenia, Dentaria, Epigaea, Erythronium, Fagus, Gleditsia, Hamamelis, Juglans, Medeola, Mitchella, Panax, Pinus, Podophyllum, Smilacina, Tilia, Tipularia,* and *Tsuga.*

Surely, *Apios* is a weed in cranberry fields, mayapple persists as a poisonous weed in recently cleared pastures, spiny *Aralia* is a weed tree, and pines are normal invaders in the old field succession down south, and could be included in a handbook of edible weeds. I included many weeds from Clyde Reed's excellent *COMMON WEEDS OF THE UNITED STATES,* from which we have extracted many of Regina Hughes' great public-domain line drawings.

Flashing through the WSSA list, I saw many other edible species that could

TABLE 1
Compounds Found or Expected in *Artemisia Annua*
(with price per gram from Sigma 1990 Catalog)

Compound	Price ($/g)
ADENOSINE, PL UBI	$4.15/g
ALANINE, PL UBI	7.20
GAMMA AMINO BUTYRIC ACID (GABA)	0.12
ARACHIDONIC ACID, PL UBI	111.00
ARGININE, PL UBI	20.65
ARTEANNUIN-B, PL FNF	
ARTEMISIA ALCOHOL	
ARTEMISIA KETONE, 2,200 PL	
ARTEMISIC ACID, PL	
ARTEMISIC ACID METHYL-ESTER, PL FNF	
ARTEMISININ, 100–5,000 LF FNF	
ARTEMISININ, 100–5,000 FL FNF	
ARTEMISINOL, PL FNF	
ARTEMISITINE, PL FNF	
ARTEMISYL-ACETATE, PL FNF	
ASCORBIC ACID, PL UBI	0.04
ASPARTIC ACID, PL UBI	0.89
BIOTIN, PL UBI	23.00
BORNEOL PL FNF	0.11
DELTA-CADINENE, PL FNF	
CAFFEIC-ACID, PL UBI?	2.01
CAMPESTEROL, PL UBI?	9688.00
CAMPHENE, PL FNF	
CAMPHOR, PL FNF	0.09
CAPILLENE, 80 PL FNF	
ALPHA-CAROTENE, PL UBI?	4240.00
BETA-CAROTENE	2144.00
CARYOPHYLLENE, PL FNF	4.32
BETA-CARYOPHYLLENE, PL FNF	
GAMMA-CARYOPHYLLENE, PL FNF	
CARYOPHYLLENE OXIDE, PL FNF	
(+)-CATECHIN, PL UBI?	5.80
CATECHOL, PL UBI?	0.03
CELLULOSE, PL UBI	0.03
CHLOROPHYLL A, PL UBI	17,700.00
CHLOROPHYLL B, PL UBI	20,555.00
CHOLINE, PL UBI	0.02
1,8-CINEOLE, PL FNF	0.09
CINNAMIC ACID, PL UBI?	353.60
CITRIC ACID, PL UBI	0.02
COUMARIC ACID, PL UBI?	1.13
COUMARIN, PL FNF	0.07
CUMINALDEHYDE, LF FNF	
o-CYMENE ?	
p-CYMENE, PL FNF	
L-CYSTEINE, PL UBI	0.19
L-CYSTINE, PL UBI?	0.12

TABLE 1 (continued)
Compounds Found or Expected in *Artemisia Annua*
(with price per gram from Sigma 1990 Catalog)

Compound	Price ($/g)
BETA-FARNESENE, PL FNF	
FARNESOL, PL ?	0.59
FOLIC-ACID, PL UBI?	4.70
FRUCTOSE, PL UBI	0.01
FUMARIC ACID, PL UBI?	0.01
GALACTOSE, PL UBI	0.22
GLUCOSE, PL UBI	0.28
GLUTAMIC ACID, PL UBI	0.02
GLUTAMINE, PL UBI?	0.14
GLUTATHIONE, PL UBI	1.43
GLYCINE, PL UBI	0.02
HISTIDINE, PL UBI	0.20
HUMULENE, PL FNF	17.20
ISOARTEMISIA ACETATE, PL FNF	
ISOARTEMISIA ALCOHOL, PL FNF	
ISOARTEMISIA KETONE, 3,400 PL FNF	
LINOLEIC ACID, PL UBI	6.60
LINOLENIC ACID, PL UBI	26.30
LYSINE, PL UBI	1.40
MALIC ACID, PL UBI?	0.01
MENTHOL, PL	0.14
METHANOL, PL UBI?	0.01
METHYL ACETATE	
3-METHYL-PINOCARVONE	
MYRCENOL	
MYRISTIC ACID, PL UBI	0.07
NIACIN, PL UBI	0.02
NICOTINIC ACID, PL UBI	0.02
OCIMENE	
OLEIC ACID, PL UBI	6.30
PALMITIC ACID, PL UBI	0.23
PALMITOLEIC ACID, PL UBI	33.65
PANTOTHENIC ACID, PL UBI	0.12
PECTIN, PL UBI	0.12
PHYTIC ACID, PL UBI?	2.16
ALPHA-PINENE	0.27
BETA-PINENE	
PINOCARVEOL	
PINOCARVYL ACETATE	
PROLINE, PL UBI	0.27
PROTOCATECHUIC ACID, PL UBI?	0.54
PYRIDOXINE, PL UBI?	0.88

TABLE 1 (continued)
Compounds Found or Expected in *Artemisia Annua*
(with price per gram from Sigma 1990 Catalog)

Compound	Price ($/g)
QINGHAOSU-IV, PL FNF	
QUERCETAGETIN-6,7,3',4'-TETRAMETHYLETHER, PL FNF	
QUERCETAGETIN-6,7,4'-TRIMETHYLETHER, PL FNF	
QUERCETIN, PL?	0.38
QUINIC ACID, PL UBI?	0.66
RHAMNOSE, PL UBI	1.00
RIBOFLAVIN, PL UBI	0.13
RIBOSE, PL UBI?	0.38
SCOPOLETIN, PL FNF	215.50
BETA-SELINENE	
SERINE, PL UBI?	0.14
SHIKIMIC ACID, PL UBI	31.75
BETA-SITOSTEROL, PL UBI	1,713.00
STEARIC ACID, PL UBI?	5.00
STIGMASTEROL, PL UBI	3.50
SUCCINIC ACID, PL UBI	0.02
SUCROSE, PL UBI	0.02
TANNIC ACID, PL UBI?	0.03
TARTARIC ACID, PL UBI?	0.03
TERPINEN-4-OL	
6,7,3',4'-TETRA-*O*-METHYL-QUERCETAGETIN, LF FNF	
THIAMIN, PL UBI	0.11
ALPHA-TOCOPHEROL, PL UBI?	0.17
3,3,6-TRIMETHYL-HEPTA-1,5-DIEN-4-OL	
TRYPTOPHAN, PL UBI	0.20
TYROSINE, PL UBI	0.09
UBIQUINONE, PL UBI?	90.10
VALINE, PL UBI	0.07
XANTHOPHYLL, PL UBI?	27,620.00
YLANGENE, PL FNF	

have been included as well: *Alisma, Anthriscus, Berberis, Campanula, Carduus, Cassia, Cenchrus, Centella, Commelina, Claytonia, Eclipta, Elaeagnus, Eleusine, Epilobium, Galinsoga, Hedeoma, Juniperus, Lactuca, Marrubium, Medicago, Melissa, Mentha, Monarda, Nepeta, Opuntia, Origanum, Passiflora, Perilla, Phragmites, Pluchea, Satureja, Tradescantia, Tussilago,* so I took some of these to bring my generic treatment in the book to 100 genera. While some of the resultant species reside in genera with several other edible species, I am restricting my account to 100 genera, each with at least one edible weed species.

Larry Mitich, past president of WSSA, has over the years published fascinating articles in his Intriguing World of Weeds series, lately found in the journal *WEED TECHNOLOGY,* hereafter abbreviated WT. I have drawn freely on his series, especially for those weeds for which I had trouble generating enough data to fill the allotted page. Likewise, when I didn't have enough information on edibility to fill the page, I drew upon the interesting folk data that abound for many weeds.

I attempted to describe the plants with the simplest terminology possible, but I am not pleased with the results. If a picture is worth a thousand words, this book is worth 100,000 descriptive words alone. Consistently, in the descriptive paragraph, I have begun with a sentence describing the overall habit of the plant, then LEAVES, then FLOWERS, and then FRUITS.

After the ***DESCRIPTION,*** there is a ***DISTRIBUTION*** paragraph. Here I have first presented some ecological information about the types of vegetation or agricultural situations in which we might expect the weed. Following is a phenological summary, a bit Ptolemaic, suggesting roughly when one might expect flowering and fruiting. Then I have rather systematically, with post-office state abbreviations, listed at least the northwestern, the northeastern, the southwestern, and the southeastern state, in that order, to circumscribe the area where one might expect to find the weed. Thus a weed ranging over all 48 contiguous states would get the following notation, WA-ME-CA-FL, meaning from Washington east to Maine and south to California, east to Florida. This is a frequent distribution pattern for many ubiquitous weeds.

This Northwest-Northeast-Southwest-Southeast presentation is a departure from that usually found in the botany books, but since most of us in the U.S. read left to right and top to bottom, such a presentation might be more appealing to the amateur. Though the weed may not be present, at least commonly so, in all intermediate states circumscribed by our mental box, NW-NE-SW-SE, their aggressiveness predicts that the weeds will occupy suitable habitats in intervening areas when the opportunity arises.

The two-letter abbreviations for the 48 states appear in the USDA Plant Hardiness Zone Map (HZM) (Figure 1). Since I could not use NE for both New England, and Nebraska, I have used MA to imply specifically Massachusetts or generically New England. Several species reach New England but not Maine. On the map, you will see the definitions of the 11 USDA Hardiness Zones defined by the average coldest nights of the year. The distribution of many perennial species, weeds and endangered species alike, seems to be determined more by extremes of temperature or precipitation than by annual means. While global warming *may* be gradually increasing the annual mean temperatures in a given site, the coldest nights in winter may be trending colder, and the hottest and driest spells may be trending hotter and drier. This augurs well for weeds, noted for their aggressiveness and tolerance of extremes, and bodes badly for sensitive endangered species. With this scenario, diversity might well decrease in temperate North America, as we lose endangered species, whose niches may well become occupied by weedier cosmopolites. As the final part of ***DISTRIBUTION,*** I have estimated the range

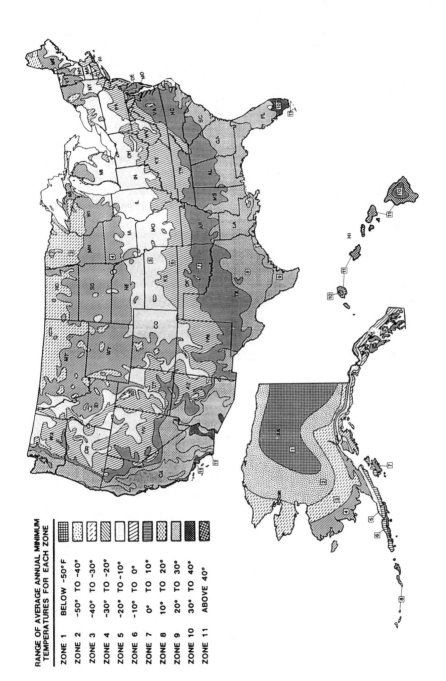

FIGURE 1. USDA Plant Hardiness Zone Map.

of hardiness zones we might expect for each weed, based on my literature and travels, and on the HZM. Perennial weeds may often have narrower ranges than annual weeds that can complete their life cycles in a 60-day growing season. Thus perennials can be better indicators of year-round climatic conditions than annuals (Duke, 1976a). Readers are reminded that the hardiness zone ranges are strictly my estimates.

FORMAT: Following the Scientific Name (in boldface uppercase and italicized) is the Family Name (in uppercase and parentheses) and common name(s). Underlined common names are those preferred by the WSSA (*COMPOSITE LIST OF WEEDS*). They tend to lowercase common names, however. They also tend to unite common names, e.g., eveningprimrose instead of Evening Primrose. Though some of their common names are contrived by translating the scientific name, their list is used by the more than 3000 members of the WSSA. In this field book, common names are separated from family name by ---. Thus an entry is as follows:

FORMAT

ACORUS CALAMUS L.   (ARACEAE) --- Calamus, Sweetflag

DESCRIPTION: Habit: LEAVES: FLOWERS: FRUITS.

DISTRIBUTION: Vegetation type; phenology; (*species*) geography (with P.O. abbreviations) and Hardiness Zones, estimated coldest to hottest.

UTILITY: Edibility; toxicity; biomass or yields; important medicinal or other uses. CAUTION: Notes on toxicity or warnings.

Alphabetical Listing of Plants

ACER SPP. (ACERACEAE) --- Maples

DESCRIPTION: Trees, often with broad rounded crowns, the bark smooth when young, flaking or furrowed in age. LEAVES: two, on opposite sides of the stem, 2–6 inches long, 2–6 inches wide, palmately lobed (i.e. the lobes and veins arising from a single basal point), with 3–9 lobes, the lobes with few to many teeth; leaf blades (and stalks) smooth and hairless; leafstalks long, sometimes as long as the blade; leaves often brilliantly colored in fall. FLOWERS: emerging with or before the leaves in spring, small, greenish-yellow or sometimes flushed with red, in few-flowered flat-topped clusters, erect or drooping; male and female flowers in separate clusters, on the same or different trees; sepals usually 5; petals 0 or 5; stamens 3–12; ovary free of the sepal tips, 2-celled. FRUITS: a characteristic pair of winged seed, on long drooping stalks, the fruits floating to earth, rotating like a helicopter.

DISTRIBUTION: Characteristic trees of the Eastern Deciduous Forest, some species in dry forests, more in moist forests, and some in swamp forests, flowering as early as February farther south; on into June in the north or high mountains, fruits ripening in autumn. Can be a weed in woods-grown ginseng or goldenseal. WA-ME-CA-FL. Zones 3–7.

UTILITY: Celebrated as the sources of maple syrups, maples have provided their seeds and inner bark as primitive foodstuffs. Some people talk about the seeds as food, others as medicine, while one MD considers them toxic. So far, I have not encountered a maple seed I enjoyed, raw, boiled, or fried. Facciola mentions what must have been an acquired taste: Indians cooking dewinged seeds in milk and butter. My horses stripped the inner bark off the maples in early spring, attracted by the upward flow of that dilute sugar solution we call maple sap. At the ripe old age of 55, I tired of not having personally experienced the tapping of the maple. So before the red maple buds had swollen in February, I bored a hole into a trunk, plugged it with a hollow reed tilted to drain into a coffee can hanging off the trunk. On good days (following cold nights), the can would fill up with a sap so watery that it had no hint of sweetness. Hours of boiling reduced it to a thin, slightly sweet solution with which I could make instant coffee, presweetened, by adding the powdered coffee to the "syrup". It takes a long time to boil down 40 quarts of sap to make the 1 quart of maple syrup. If you have a fixed campsite and permanent outdoor fire, you may find the boiling-down process practical. The sap can be concentrated by leaving it out at night, throwing off the ice (relatively pure water) the next morning, and repeating the process as long as the nights are freezing. WSSA lists only *Acer macrophyllum* Pursh

Acer saccharum Marsh.

(Bigleaf Maple), *A. rubrum* L. (Red Maple), and *A. saccharum* Marsh (Sugar Maple) as weedy maples. Many people consider silver maple (*A. saccharinum* L.) equally weedy. With shallow, wiry surface roots, maples can be serious weeds in woods-grown ginseng, interfering seriously with cultivation. CAUTION: Dr. John Churchill, in a well-documented Smithsonian lecture, reported toxic properties from some maple seeds.

ACORUS CALAMUS L.   (ARACEAE) --- Calamus, Sweetflag

DESCRIPTION: Aromatic, perennial herbs to 6 feet tall, often in dense, pure stands, the rather woody rootstocks often exposed at the surface of the ground. LEAVES: irislike, sword-shaped, pungently aromatic, 6–36(–72) inches long, flat, with smooth, toothless margins and parallel veins. FLOW-ERS: in club-shaped clusters emerging at angles from a greenish stalk, sugges-tive of the leaf but triangular rather than flat in cross section; the flowers minute, surrounding the core of the clublike axis, each flower with 6 sepals, 6 stamens, and a 2- to 3-lobed ovary. FRUIT: a gelatinous berry (rarely sets seed in the U.S.).

DISTRIBUTION: Often forming pure colonies in seepage slopes, sunny swamps, or at the edges of water bodies, more common on the coastal plain than in the piedmont or mountains, flowering from May to June in the Car-olinas, May to July in Maryland. Could be a weed with aquatic crops. WA-ME-CA-FL. Zones 4–8.

UTILITY: The clambering stems or rootstocks are one of Father Nature's natural candies, ready for limited consumption right out of the swamp (if the swamp is clean, a big IF; recently we were told that even in pristine primitive areas, the water may harbor *Giardia*). Some authors suggest stimulation, others even hallucination, from overconsumption. In the Depression years, sweet flag was chewed in lieu of tobacco, some claiming it cut the desire to smoke. In *HERBAL BOUNTY,* Steven Foster describes nibbling on the root to keep himself awake on long drives. More properly, the tough rootstocks can be boiled with maple sap for days to prepare candied sweet flag. With their soapy, yet gingery and peppery taste, fresh roots will not please all comers. Facciola (1990) mentions that inner portions of young shoots make good salad material. Penobscot Indians hung the roots around the house to keep away disease; now the plant has been proven to contain insect-repellant and antiseptic compounds. Ojibwa soaked their gill nets in the tea, containing asarone, to attract fish. Interestingly, Panamanians used fruits of an unrelated pepper species, containing safrole, close kin to asarone, as fish bait. Sassafras roots, also containing safrole, were similarly used by North American Indians. Commercial plantings have produced a ton of root per acre. CAUTION: Foragers graze at their own risk when they eat the tender young shoots in spring. Since some strains of sweet flag in India, containing beta-asarone, proved carcinogenic in animals, the U.S. Food and Drug Administration (FDA) curtailed the use of sweet flag and its extracts in human foods. Beta-asarone is rather similar to safrole, the alleged carcinogen in sassafras. (Studies

Acorus calamus L.

in *Science* magazine, April 17, 1987, suggest though that a 12-ounce can of sassafras root beer is less than one tenth as carcinogenic for its safrole as a 12-ounce can of beer for its ethanol. Perhaps, because of the lack of a lobby, the root beer was banned while the alcoholic beer still thrives.)

AGROPYRON REPENS L.   (POACEAE) --- Couchgrass, Devil's Grass, Doggrass, Quackgrass, Quickgrass, Scotch Quelch, Twitchgrass, Witchgrass

DESCRIPTION: Perennial grass, the stems waxy gray to bluish, hairless, to 3 to 3.5 feet tall, from a creeping whitish or yellowish rootstock, up to 3 feet or more long. LEAVES: one at the node, alternating or spirally arranged along the stem, the blades to 12 inches long, 1/2 inch wide, sometimes slightly hairy on the upper surface, with no obvious leafstalk and no teeth; veins parallel. FLOWERS: in tight, terminal clusters (spikes) to 10 inches long above the leaves, flattened in a zigzag fashion against the central stalk, with short bristles extending beyond the individual florets. FRUITS: minute grains in ribbed green enclosures terminating in small bristles.

DISTRIBUTION: Ubiquitous weedy grass, occurring in old fields, feedlots, orchards, and poorly maintained pastures, roadsides, and waste places. Flowering mostly from June, fruiting into August, at least. Can be a serious weed in beans, beets, cereals, corn, and tobacco. WA-ME-CA-SC. Zones 3–7.

UTILITY: Famous and obnoxious for their tendency to grow right through a potato, the horizontal edible rootstocks, sometimes chewed like licorice, can be scorched and used as a coffee substitute, or ground up and made into beer. Seeds, though small, are produced at rates up to 400 per stalk, and can be used for making breadstuffs and beers. As with other members of the grass family (POACEAE), the plant can be used for pasturage or cut for hay. Reported hay yields are 2–3 tons/acre. Rootstock yields may be as much as 2–2.5 tons/acre, 10% of which may be mucilage. Quackgrass is said to have enough nitrogen to make it useful fodder, not high enough to induce nitrate or nitrite poisonings. The tough "couch" of interlocked rootlets makes quackgrass useful in stabilizing slopes, and is used to stabilize new dikes in Holland. Michigan studies suggest that quackgrass is effective at reclaiming nutrients from sewage effluent sprays. Although foragers use the rule of thumb that all grasses are edible, wilted grasses sometimes contain dangerous levels of cyanide, and in wetter climates, tainted grains may cause ergot poisoning. Pollen can cause hay fever. Most famous as an herbal diuretic, quackgrass is also folklorically considered useful in cancer, cystitis, enuresis, gallbladder ailments, gout, gravel, rheumatism, scleroses and worms. A leading Greek physician of the 1st century A.D., Dioscorides, recommended the root for gallstones. Duke and Wain (1981) cite quackgrass as astringent, demulcent, depurative, diuretic, emollient, and sudorific. Methanol extracts can be used to control mosquito larvae (WT1:184.1987). Farmers' claims that quackgrass can damage crops have been justified by the recent discovery of herbicidal compounds in the quackgrass. *Science News* (March 16, 1991) reports the

Agropyron repens L.

discovery in the extracts of a molluscicide that controls slugs (slugicide), even controlling *Arion subfuscus,* potentially the most serious slug pest in the U.S. WSSA lists 3 weedy species, under the generic name *Elytrigia.*

ALISMA SPP.   (ALISMACEAE) --- Mad-Dog Weed, Mud Plantain, Waterplantain

DESCRIPTION: Low herbaceous perennial aquatic herbs, the stalked leaves all arising from a basal whorl (rosette). LEAVES: egg-shaped, broadest at, above, or below the middle, 1–8 inches long, 1–6 inches wide, pointed (acute) at the tip, rounded or notched at the base where attached to the leafstalk, toothless on the edges (entire), with usually 5–7 strong veins arising from the base, the laterals arching out and extending to the tip, connected by prominent cross nerves as well; stalks about as long as the leaf blades. FLOW-ERS: in diffuse clusters, rising from among the basal leafstalks, the flowers in whorls on branching stalks; sepals 3; petals 3, pinkish or whitish; stamens 6–9; ovaries several. FRUITS: minute, greenish, seedlike, in clusters resembling small green blackberries.

DISTRIBUTION: Marshes, mudflats, ponds, and slow sluggish waters, flowering from April (in the Carolinas) or June (Maryland) until frost. Strictly an aquatic weed; rarely a weed in groundnut which itself is a weed in cranberry. WA-ME-CA-FL. Zones 3–8.

UTILITY: I maintain that the grazer or survivalist is better off near streams, not only for the animal bounty (such as crayfish, fish, and frogs), but for the herbal bounty, and that most important survival item, water itself. In survival situations in the tropics, death from dehydration is more likely than death from starvation, especially among the obese. Like so many marsh or aquatic plants, the waterplantain has a starchy, swollen, or bulblike base of the leafstalk, which can be rendered edible, if not enjoyable, by washing and drying to remove acridity. Those that I have eaten fresh and raw suggest a slightly bitter raw potato in taste. Wading down reasonably clean streams near my home with foraging classes, I have enjoyed the leaf bases, merely washing them free of mud and then eating, with no apparent ill effects. But there were several other edible species in the herbal bounty along that beautiful stream. Paradoxically, some orientals believe the plant stimulates the female genitalia and promotes conception, while the seeds are believed to promote sterility. Roots, famed as a diuretic, are used by orientals for abdominal swellings, ascites, beri-beri, bladder distension, chyluria, diabetes, dropsy, dysuria, edema, fever, gonorrhea, painful micturition, and polyuria. Thus it might be useful in diabetes. Cherokee poulticed the roots onto bruises, sores, swellings, ulcers, and wounds. Russians, apparently without scientific justification, regard the plant as a specific for hydrophobia, hence the name mad-dog weed. CAUTION: One always risks parasitic infections when ingesting uncooked aquatic plants. While I have ingested these plants with no apparent problem, some authors describe them as poisonous. Grieve (1974)

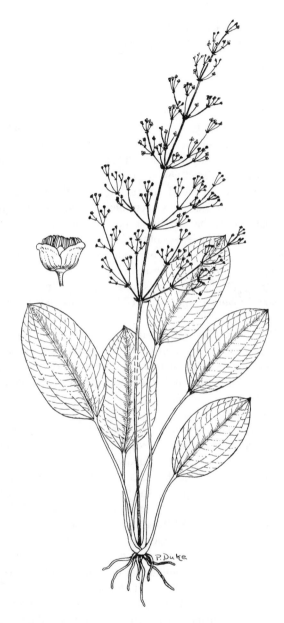

Alisma subcordatum Raf.

adds that bruised leaves are rubefacient and may even blister the skin. I have not experienced this in my dealings with the plant. Grieve adds that they are injurious to cattle. They could be poisonous to you. WSSA lists 3 *Alisma* species as weeds.

ALLIARIA PETIOLATA (M. Bieb.) Cavara and Grande (BRASSICACEAE) --- Garlic Mustard, Jack-by-the-Hedge, Sauce Alone

DESCRIPTION: Low, annual or biennial herbs, often emitting the aroma of garlic if crushed. LEAVES: as a winter annual (a plant that germinates and establishes itself in summer or fall of one year, persisting overwinter, flowering and dying the following year), the garlic mustard has nearly circular winter rosette leaves (that's all you see in December in Maryland), rounded to heart-shaped with coarse, round, irregular teeth on the margins, 2–8 inches broad, about as long, the leafstalks even longer, with 3–9 veins fanning out where the leafstalk joins the leaf; in spring, the plant, like radishes, bolts (sends up a flower stalk), the leaves of the flowering stalk gradually becoming smaller, narrower, and more triangular, with proportionately shorter stalks. FLOWERS: small, ca. 1/4–1/3 inch across, in few-flowered clusters at the end of this central stalk, the lowermost opening first; sepals 4, green; petals 4, white; stamens 6; ovary 1, simple. FRUIT: a narrow pod, 1–2 inches long.

DISTRIBUTION: Here and there common in deep, but rather open forests, such as the alluvial forests in Carolina and Maryland; also in old fields and along roadsides. Flowering April to June, fruiting and dying back May to July. A weed mostly of azalea and wildflower gardens in partial shade. MN-MA-KS-NC. Zones 3–6.

UTILITY: A year-round potherb, this is one of my few Christmas greens, to accompany the more substantial December foods like groundnuts, putty-roots, and walnuts. I have eaten all aboveground parts of the plant, leaf, stem, flower, and fruit, and find them, as with other members of the mustard family (BRASSICACEAE), hot and spicy raw, improved by cooking with a couple changes of water. Low-income Europeans once used the leaves on bread to make sandwiches called "sauce alone", and they added garlic mustard leaves to spice up lettuce, mutton, pork, and salt-fish. Lacking the real condiments, I would add garlic mustard to any recipe calling for garlic or onion. Brill (n.d.) shares my opinion that the root is edible before the plant bolts. Recently, the NCI hinted that, if ingested, members of the mustard family (such as cabbage, cauliflower, broccoli, collards, kohlrabi, brussels sprouts, turnip greens, mustard, watercress, etc.) may reduce the incidence of cancer. In its new Designer Food Program, the NCI is including this weed for study because it embraces the chemistry of both the mustard and the garlic, both with well-deserved chemopreventive reputations; e.g., it does contain isothiocyanates, one of the cancer preventives of the mustard family, and allyl sulfides, cancer preventives of the garlic family. Zennie and Ogzewalla (1977) showed that the rosette leaves contain twice the chemopreventive beta-

College of St. Francis Library
Joliet, Illinois

Alliaria petiolata (M. Bieb.) Cavara & Grande

carotene of spinach. CAUTION: Don't eat a supposed garlic mustard if you don't get a garliclike odor when you crush it; otherwise you may have the leaves of look-alike squaw-root (*Senecio*), a widely used Amerindian medicinal plant which may contain carcinogens (pyrrolizidine alkaloids). It can also be confused with *Glechoma,* if you don't have a good sense of smell.

151,426

ALLIUM VINEALE L.   (LILIACEAE) --- Field Garlic, Wild Garlic

DESCRIPTION: Wiry, onionlike herbaceous perennials, arising from a white, underground bulb comprised of "cloves" (sections) like garlic, the entire plant reeking of garlic. LEAVES: early leaves basal, round, and hollow, 4–12 inches long, striate, the later leaves long and narrow, clasping tightly below, around the circular green stem, folding out and falling loosely above. FLOWERS: in flat to rounded clusters at the tip of the stem, greenish, reddish, or purple, at first enclosed in a green pouch; sepals 3; petals 3, similar to the sepals; stamens 6; ovary 3-celled, free of the petals and sepals. FRUIT (rarely setting in the U.S., often replaced by plantlets or bulblets): a 3-lobed pod, with about 6 black seeds.

DISTRIBUTION: Introduced weed, aggressive in lawns, old fields, disturbed areas, and forest edges, and a nuisance in winter cereal crops; flowering May to July, fruiting (or new plants flowering) July to frost. WA-MA-CA-FL. Zones 5–9.

UTILITY: Indians used wild garlic in cookery, and dried it as a winter condiment. I drink the pot liquor (really an herb tea) of wild garlic with garlic mustard, apparently repelling vampires, as true garlic was alleged to do. From a savory point of view, I enjoy the pot liquor, but not the greens or "chives" (chopped leaves). I have endured wild garlic cloves boiled as a vegetable and added, like chopped onion, to vegetables, but they leave a bad aftertaste. I find the flower heads, enclosed within their pouches (before they open), much milder than the rest of the wild garlic. Milk produced by cattle grazing on field garlic can be tainted with the garlic aroma. I have added wild garlic cloves to martinis, but don't think either was improved thereby. Colorado Indians added it to their buffalo meat (Kindscher, 1987), much as I add onion to hot dogs and hamburgers. In a survival situation, were I seeking something antiatherosclerotic, antiseptic, candidicidal, fungicidal, hypoglycemic, and hypotensive like the onion or garlic, I would not hesitate to use field garlic. I used garlic cloves successfully with an abscessed ear once when no doctor could see me and the infection threatened to spread to my eyes as well, which almost were swollen shut. A 1990 symposium on the "tame" garlic, *Allium sativum,* suggested that garlic could ward off heart disease, America's number one killer, e.g., by thinning the blood (slowing down its tendency to clot) and inhibiting the synthesis of cholesterol or lowering blood levels of cholesterol. Garlic was also shown to play a role in the reduction of the tendency towards breast cancer and high blood pressure. CAUTION: Although most foraging books treat wild garlic as edible, others suggest it may be poisonous; even onions and garlic have proved poisonous to animals when consumed in

Allium vineale L.

large quantities. Some people are seriously allergic to onions and many find them relatively indigestible. Amateurs best be wary. They might confuse the dangerous star-of-Bethlehem or much more dangerous death camas with the wild onion. If it doesn't smell of onion or garlic, beware. WSSA lists the Wild Onion, *Allium canadense* L. as a weed also.

AMARANTHUS SPP.   (AMARANTHACEAE) ---
Pigweed, Redroot

DESCRIPTION: Coarse, annual, often weedy herbs, sometimes 8 or 9 feet tall. LEAVES: one at the node, usually alternating along the stem, 1–10 inches long, 0.5–4 inches broad, willow-shaped to egg-shaped, usually broadest below the middle, tapered at the tip, tapered or rounded at the base, marginally toothless (entire), with a long leafstalk. FLOWERS: unisexual (male or female, with some occasionally hermaphrodite), in clusters at the tips of the stems, often with smaller clusters arising from the angles of the upper leafstalks with the stems; sepals 3–5, with no petals differentiated, usually greenish, sometimes pinkish or reddish; stamens 2, 3, or 5 in the male flowers; ovary 1-celled, with 2, rarely 3, processes (styles) emerging from the tip (apex). FRUIT: a minute, green, saclike structure (utricle), turning brown, with 1 brown or black seed. There may be hundreds of these tiny fruits in the more complex flower clusters.

DISTRIBUTION: Often cohabiting with the equally weedy lambsquarter (*Chenopodium*), the pigweeds are more inclined to flourish in the heat of summer, the lambsquarter in spring and fall. Flowering from June to August, fruiting to frost. Common in annual crops, sometimes producing more biomass than the crop, also in old fields, pastures, and waste places. *Amaranthus retroflexus* (Redroot Pigweed): WA-ME-CA-FL. Zones 4–9.

UTILITY: Young plants, especially more tender parts, are widely available as potherbs in late spring and early summer, following the lambsquarter. I have eaten the Maryland species, enjoying the potherb as much as spinach. (But this is a personal taste; I don't like spinach, lambsquarter, chard, or beet tops, all members of the spinach family, nearly as much as I like potherbs of the cabbage family. I even speculate that blacks and southerners as a rule prefer potherbs of the cabbage family, while northerners tend to prefer potherbs of the spinach family.) I have eaten boiled flower clusters (inflorescences) of the spiny pigweed and find them as edible as those of the other pigweed species, like a slightly acrid spinach. Seeds of the various pigweeds can be gathered in the milk stage (green) or ripe as a cereal, and then ground up, with or without the fibrous greenery surrounding them, to make a mush or breadstuff. Some Indians used to gather the plants green so as not to lose any seed to shattering. Zuni Indians stripped the seeds and accompanying chaff or prostrate pigweeds into their hands, blowing the chaff away. Seed of some species can be popped like miniature popcorn. CAUTION: Hungry animals, including hardcore foragers, could possibly ingest too much green amaranth (or lambsquarter or spinach, for that matter) and induce nitrite or nitrate poisoning — but it would probably take pounds of the stuff. Could be that

Amaranthus retroflexus L.

Popeye's elbow sticks out so because of too much nitrate or nitrite in his spinach, messing up his uric acid metabolism, and causing gout of the elbow. All things in moderation! WSSA lists 15 weedy *Amaranthus,* any of which might be sampled judiciously.

AMELANCHIER SPP. (ROSACEAE) --- Juneberry, Serviceberry, Shadbush

DESCRIPTION: Deciduous trees or shrubs, sometimes in colonies, flowering as the leaves unfold in spring. LEAVES: one at the node, alternating on the stem, narrowly to broadly egg-shaped, broader at, above, or below the middle, 1–4 inches long, 1/2–2 inches broad, with numerous fine teeth along the edges, and numerous veins feathering off the middle vein (midrib); blades rounded or pointed at the tip; tapered, rounded, or notched at the base where it merges with the leafstalk. FLOWERS: in sparsely flowered clusters at the stem tips, emerging and drooping from the angle formed by the leafstalk and stem; sepals 5, united in a tube, the pointed lobes alternating with the 5 white or pinkish petals; stamens ca. 15–20; ovary 5-celled, surrounded by the green sepal tube, terminating in 5 processes (styles). FRUIT: a rounded, reddish, purplish to blackish berry, capped at the tip with the 5 sepal lobes, rather like a miniature apple (it belongs to the apple family).

DISTRIBUTION: Occurring in deciduous forests or at the edges of coniferous forests or savannas, the several species tend to hybridize, making their identification quite difficult at times. Reported from lowland to upland swamps and balds, the shrubs flower in spring, the fruits rarely persisting to winter, being very attractive to birds. *Amelanchier arborea*: MN-MA-OK-FL. Zones 4–8.

UTILITY: Among the earlier fruiting plants in Maryland, the amelanchiers bear fruits that are pleasant tasting, whether eaten out of hand or processed into jams, jellies, or pies. Fruits of those species I have sampled are good, sometimes juicy, sometimes dry, with flavors between the drier rose hips and the wetter plums. Indians used these fruits like blueberries and cranberries, dried and mixed with suet or beef to make a long-lasting pemmican. The blueberries and cranberries at least were shown later to contain antioxidants which would slow down the spoilage of the pemmican. Naegele (1980) notes that Cheyenne made a tea from the leaves. Some Indians lined vats with spruce bark, covering it with amelanchier fruits, then red-hot stone, then another layer of berries, then hot stone, until the vat was full, letting steam about 6 hours. Then the berries were removed, crushed, dried over a slow fire, and consumed on the spot or much later, as we today consume more familiar dried fruits. In a very interesting account of *A. alnifolia,* Kindscher notes how Blackfoot Indians made sausage from animal fat and serviceberries. Berries were placed on a perforate skin, elevated out of the reach of children and animals. When dry, they were mixed with fat and stored in fetal deer or

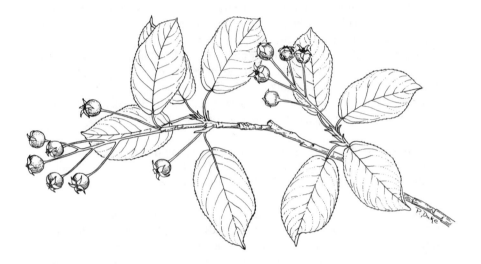

Amelanchier laevis Wieg.

sheep hides. The pits or stones, like those of other members of the rose family, contain cyanidelike compounds, giving off the pleasant benzaldehyde flavor of bitter almonds. Dried leaves also may have these cyanidelike compounds; still, they can be dried and used in herb teas. CAUTION: In large quantities, the pits and leaves could cause cyanide poisoning. NOTE: Only *Amelanchier laevis,* the <u>Allegheny Serviceberry</u>, is listed as a weed by WSSA.

AMPHICARPAEA BRACTEATA (L.) Fernald
(FABACEAE) --- Hog Peanut

DESCRIPTION: Delicate herbaceous twining annual vines, climbing over other plants as a rule. LEAVES: cloverlike with 3 leaflets (trifoliolate), alternating on the stem, the leaflets 1/2–4 inches long, egg-shaped, broadest near the rounded base, slightly pointed at the other end, with lateral nerves coming off the middle vein like distantly spaced plumes of a feather (penninerved), the stalk of the compound leaves about as long as the leaf. FLOWERS: in few-flowered, stalked, drooping clusters emerging from the angle of the leafstalk with the stem; white or purplish; sepals 4, united to form a tube from which emerge the pealike petals; stamens 10; ovary simple. FRUIT: a beanlike pod with 1–4 beanlike seeds. In addition to the typical flowers described above, rudimentary flowers near the base of the plant are pushed into the ground like a peanut, containing only 1 edible seed each.

DISTRIBUTION: Quite common in damp, sunny meadows and in moist, shaded forests, flowering, especially in sunnier situations, July to September, fruiting August to frost. MT-MA-TX-FL. Zones 3–8.

UTILITY: The vine is unusual in forming both aerial and subterranean pods, the above-ground pods with 1–3 tiny seeds, the subterranean with only 1 fleshy succulent seed. Both, especially the latter, are edible raw or cooked, but they don't offer much sustenance. In fall, one locates the subterranean seeds by tracking the vine back to the ground, then scooping up a 2-inch hemisphere there, carefully sifting out the 1 or 2 seeds that may be waiting. These underground seeds look a lot like swollen blood-filled ticks. Apparently, mice and voles, more adept than humans, are better at finding the hog "peanuts". This is documented, at least in the Dakotas, where the Indians once raided mouse nests for their caches of hog peanuts, leaving the mice some Indian corn in exchange. Chewy raw, like a raw kidney bean in flavor and texture, the seeds are tastier cooked, boiled, or fried. I've ground both aerial and subterranean seeds and added them to prefab cornbread mix to make delicious muffins, suggestive of old southern cracklin' bread. One writer describes them as the best of vegetables used in stews (see Kindscher). Like other members of the bean family, these too can induce gas, or flatus. One herbalist, using beans for her vegetarian protein, told me that she was pretreating the beans with peroxide, rendering them less gas-inducing. Then I read that Mexicans add epazote (*Chenopodium ambrosioides*) for the same purpose. Epazote, like *Artemisia annua,* is one of the few plants containing endoperoxides. Peroxides release oxygen in anaerobic environments like the rumen of cattle, perhaps causing evolution of less methane and proportionately more carbon dioxide, these being our two main greenhouse gases. If we

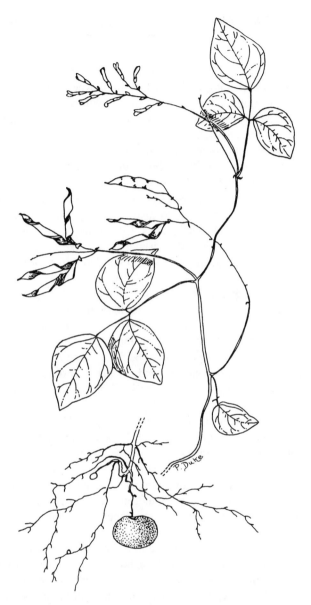

Amphicarpaea bracteata (L.) Fernald

replaced ruminant animal protein in our diet with bean protein, we could conceivably alter the global warming trend, perhaps significantly. WSSA does not list this as a weed, although it can be quite weedy.

APIOS AMERICANA Medik.   (FABACEAE) --- Groundnut

DESCRIPTION: Climbing herbaceous vine twining around other plants for support, sometimes exuding a little white milk where broken. LEAVES: alternating on the stem, pinnately compound, with 5–7, rarely 3–9, egg-shaped leaflets, mostly broadest below the middle, somewhat pointed at the tip, untoothed on the edges, rounded at the base, the leaflets individually almost stalkless, but the compound leaf itself stalked; leaflets with several veins arising from the midrib and arching out to the edge of the leaf; leaves 4–10 inches long, the individual leaflets 1–3 inches long. FLOWERS: in a tight stalked cluster, the stalk arising in the angle formed by the stem with the leafstalk, pea-shaped, purplish to maroon, rarely white, resembling diminutive wisteria flower clusters; sepals 4–5; petals 5; stamens 10; ovary 1-celled. FRUIT: a long green pod, drying brown like a string bean, 2–5 inches long, 2- to 9-seeded, the seeds small and lentil-like, brown.

DISTRIBUTION: Aquatic or semiaquatic, found in marshes and moist meadows and draws, but seeming to persist in drier situations of the garden once transplanted, flowering June to July (North Carolina to Maryland), fruiting until frost, the fruits hanging on into early winter. My groundnut comes back annually from unharvested tubers intermingled with unharvested tubers of the Jerusalem artichoke. Around my place, it often twines in alder bushes overhanging Rocky Gorge Reservoir. WY-ME-TX-FL. Zones 3–9.

UTILITY: The groundnut, introduced to the Pilgrims by the Indians, is such an important forager's food, that it may have been responsible for the Pilgrims' survival through their first winters. Without the groundnut, there might have been many more Lost Colonies, and the Caucasians might not so readily have displaced the generous Indians who taught us the groundnut. With three times the protein of the potato, this marvelous root, tasty raw or cooked, can be located by the discerning forager all winter long. The often numerous tubers are strung along the wiry thin roots like beads on a necklace. Indian women reportedly collected a half bushel of them a day, their harvest much better than any of mine. But scientists at LSU can now get several tons of groundnut per acre. Although the leaves fall off in winter, the bare brown stems, locally flattened, can still be recognized, clambering through the alders and traced carefully to the roots, attached to which are the protein-rich tuberous treasures. The roots are delicious when hash-browned in bacon drippings. Some liken the flavor to mushrooms, a more apt comparison than parsnips or prunes. Menominee Indians candied them in maple sap. Cooked seeds suggest lentils in flavor and appearance; and dry beans need to be soaked overnight, while

Apios americana Medic.

the green beans can be cooked outright. Ground, ripe seeds make an interesting addition to cornbread. Something may be breaking on the groundnut: a Japanese firm asked a friend for a quotation on 100 tons. Perhaps it has many of the cancer-preventive compounds occurring in the soybean. WSSA lists no weedy *Apios*. It is a serious cranberry weed.

ARCTIUM MINUS (Hill) Bernh. (ASTERACEAE) --- Common Burdock

DESCRIPTION: Coarse biennial or perennial herbs, as much as 6 feet tall when in flower. LEAVES: huge, in basal rosettes during the vegetative year(s), the stalks seemingly arising from the ground, hollow (or solid in the Great Burdock, *Arctium lappa*), the blades egg-shaped to nearly rounded or heart-shaped, somewhat pointed at the tip, irregularly saw-toothed at the margin, heart-shaped or notched at the base, whitish beneath; flowering stems have leaves one at a node, alternating along the stalk, the stem leaves smaller, with shorter stalks, grading into much smaller leaflike appendages below the flower clusters. FLOWERS: thistlelike, of dense clusters of minute flowers aggregated into the "thistle"; sepals 5, minute; petals 5, fused to each other throughout much of their length; stamens 5; ovary simple, with 2 terminal processes. FRUITS: aggregated in the developing burrs which readily attach to animals and foragers, explaining the wide distribution of this pesky weed.

DISTRIBUTION: Especially common in full sun or partial shade about old rural dwellings, pastures, old fields, and waste places, starting to flower in June (North Carolina) to July (Maryland) and fruiting on up to frost. Flowering and fruiting specimens are all but useless for food. As with most biennials, it is the food in the roots of the first year's winter rosette that is most copious. The plant exhausts this when it goes to flower (bolts) in its second year. WA-ME-CA-AL. Zones 4–8.

UTILITY: Cooking this malodorous herb produces something deserving the appellation "vegetable". Having read that cooking with baking soda partially breaks down the tough fibers in the leafstalks, I bravely bit into one that had survived two changes of water. With the soda, I gained some tenderness, lost some vitamins. The large roots, diced, came out quite tasty after similar treatment. Remembering the celery-onion soup mom used to give me when I had a cold or flu, I gathered up more burdock leafstalks and some wild garlic flower heads, and boiled them up together. The result: one of the better culinary episodes in the foraging chapters of my life, especially after I added salt and pepper and oleo. Rogers and Powers-Rogers (1988) note that claims of endurance and sexual virility sell a lot of this herb in Japan. Yanovsky (1936) notes that Iroquois ate the burdock (*Arctium lappa*) as a greens. Indians also used the roots in soups, as have I, and they dried and stored the roots for winter use. Scorched roots serve as coffee substitutes. Only in 1990 did I learn that the NCI had an interest in burdock lignans as chemopreventive antimutagens. Ironically burdock is conspicuous in several allegedly "quack" cancer remedies, e.g., the "Essiac" and "Hoxsey" formulas. With my immune system and liver overassaulted by overmedication, I am imbibing a

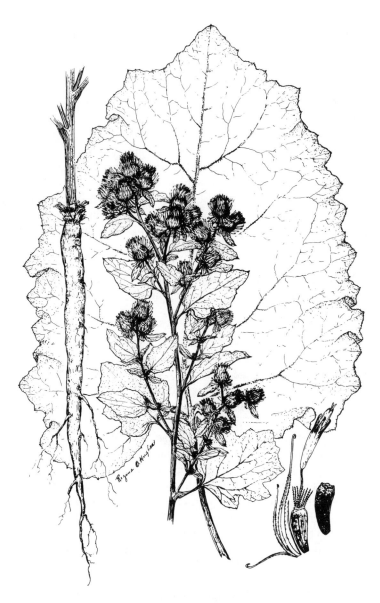

Arctium minus (Hill) Bernh.

little of the Essiac formula, just in case! Hartwell mentions folk use of *Arctium* for cancer in such diverse places as Belgium, Canada, Chile, China, India, Indiana, Italy, Japan, Oklahoma, South Africa, Spain, Ukrania, and Wisconsin. All three *Arctium* species on the WSSA list might be sampled judiciously.

ARUNDO DONAX L.   (POACEAE) --- Biblical Reed, Cane, Giant Reed

DESCRIPTION: Tall, erect, tufted or clumped bamboolike perennial grass, from a knotty horizontal rootstock; to nearly 20 feet tall, 1 inch in diameter; stems grayish green, hairless, sometimes rooting at the lower nodes. LEAVES: basal and then one at the node, alternating on opposite sides of the stem, the blades to 24 inches long and to 2.5 inches broad, rarely hairy at the junction of leaf and stem, sometimes rough to the touch (scabrous) at the edges, with no obvious leafstalk and no teeth; clasping where the leaf joins the stem; veins parallel. FLOWERS: numerous, closely spaced in plumose terminal clusters up to 25 inches long, 6 inches broad. FRUITS: small, possibly not setting regularly, obscured by the surrounding hairs.

DISTRIBUTION: Weedy wetland grass of moist roadsides and waste places, often forming extensive stands. Flowering mostly September to October in the Carolinas. CA-MD-TX-FL. Zones 5–9.

UTILITY: Only young shoots are palatable as forage for cattle and foragers. The young rootstocks and stalks are sometimes sweet enough (3–5% sucrose) to serve as sugarcane substitutes. Italians use the root infusion as a diaphoretic, diuretic, emollient, and lactifuge. Egyptians use it as diaphoretic and diuretic. Spaniards use it as diuretic and lactifuge, taking a decoction of the rhizome (1 ounce rhizome per liter of water, boiling 15 minutes). The local anesthetics known as lidocaine, lignocaine, and xylocaine were patterned after the alkaloid gramine found in *Arundo*, not cocaine as many have suggested. How well I remember my latest encounter with lidocaine, the first day of spring, 1991, when they had to give my overdrugged body not one, but three, local injections of lidocaine before injecting me with contrast dyes preparatory for a mye-logram to better define my herniated disk. By April Fools' Day, my cranium still seemed to be clogged with the contrast dye. But the lidocaine had spared me the pain on the day of the dye injection. (Mortality rates are not bad nowadays from the contrast dye injections, though odds were once posted at one fatality per 50,000 injections, much better than the 1 fatality per 4,000 injections with the disk-dissolving chymopapain). In Mallorca, they take ringlike sections of the cane from old fences (as much as 40 years old) as "sterile" applications to wounds. Boiled in wine with honey, the root was reported in the 12th century to be used as a folk remedy for "cancer of fleshy parts". Romans were reported (ca. 180 A.D.) to use reed roots for condy-lomata and indurations of the breast. Reed has been used as a cellulose source for rayon manufacture. Wherever these aquatic plants grow, one finds them used for simple construction, for fencing, trellises, roofing, basketry and mattings, and occasionally for forager foods. It is such a vigorous source of

Arundo donax L.

biomass (to 30 tons/acre) that it is often proposed as an energy source. Variegated cultivars are often grown as ornamentals. CAUTION: The alkaloid gramine, with activity similar to *d*-pseudoephedrine, is hypertensive to dogs in small doses, hypotensive in larger doses. Reportedly, it contains bufotenine and dimethyltryptamine.

ASCLEPIAS SYRIACA L. (ASCLEPIADACEAE) ---
Common Milkweed

DESCRIPTION: Upright, often unbranched herb to 6 feet tall, nearly suc-culent, exuding a copious white latex where broken. LEAVES: two on op-posite sides of the node, the position of one pair alternating with that of the next pair up the stem, nearly stalkless, 3–12 inches long, broadly oblong, elliptic, or egg-shaped, mostly broadest at or near the middle, rounded at both ends, toothless on the edges, with many veins arising featherlike from the midvein. FLOWERS: in rounded or flat-topped clusters at the tip of the stem and in the leaf junctions (axils) of the upper leaves, the stalks of the flower clusters clearly longer than the leafstalks; sepals 5, basally united; petals 5, basally united, cream and/or greenish, with hints of maroon and purplish; stamens 5, their stalks united; ovaries 2, completely separate. FRUITS: solitary or paired, large green fleshy pods to 5 inches long, finally splitting in winter to reveal numerous, flattened, brown to black seeds, each with a tuft of long hairs.

DISTRIBUTION: Abundant weed in pastures, prairies, old fields, and the like, flowering June to August, fruiting July to frost. ND-ME-OK-GA. Zones 4–8.

UTILITY: Said to be poisonous to livestock, but almost never eaten by livestock, the milkweed is rather common fare for learned foragers, living dangerously, for the milkweed contains cardiac glycosides. Still, many for-agers besides Jim Duke have enjoyed the young shoots, flower buds, and/or green fruits, boiled in two changes of water. I classify these among the better tasting wild foods, especially when seasoned with wild garlic tops and/or garlic mustard or bulbous cress. Facciola mentions the sprouts as food. In-dians, e.g., the Chippewa, added the cooked buds and flowers to meat soups or stews. June 1, 1986, found me too late for the tender young shoots, yet too early for many flower buds, so I pinched off the top two pairs of obvious leaves from a couple dozen plants to boil up a tasty batch of greens, throwing off the first change of water. Reading that the Indians used the coagulated milk from the stems as a chewing gum, I bled several stems into a tablespoon, let it coagulate for a few days, and finally proved to myself that it could in fact serve as a chewing gum. USDA scientists studied the milk as a potential source of renewable hydrocarbon fuels, and the floss for fiber. Kindscher says the floss was used to cover the pubes of maidens "who had never tasted man". CAUTION: While I have indulged in milkweed foods, I remind you that, like many plants containing a white milk, milkweeds can be poisonous if not properly prepared. Another plant very similar to the milkweed is Hemp Dogbane, *Apocynum cannabinum* L., which, like milkweed, has opposite

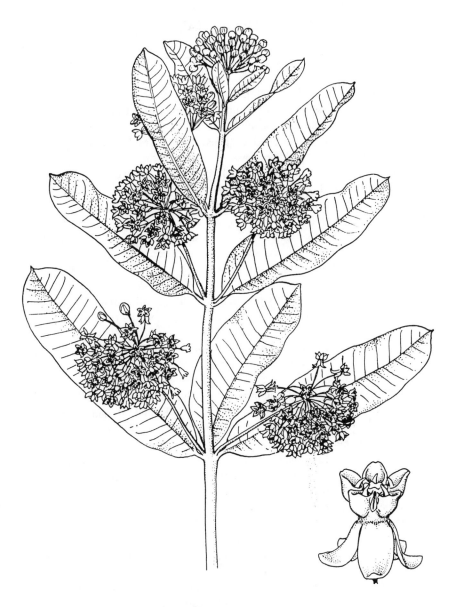

Asclepias syriaca L.

leaves and milky sap. Two of my friends have gathered the more bitter dogbane tops for milkweed shoots, with unpleasant but not fatal results. Yes, I am even a little leery of the milkweed chewing gum, even after flavoring it with peppermint. WSSA lists 12 weedy species of milkweed *Asclepias* but they can't all be recommended as food. Some could be quite dangerous.

ASIMINA TRILOBA (L.) Dunal   (ANNONACEAE) ---
Pawpaw

DESCRIPTION: Deciduous aromatic (malodorous to some) shrub or small tree to 40 feet tall, sometimes forming thickets. LEAVES: one per node, alternating on the stem, tending all to lie in the same plane, 6–12 inches long, 3–6 inches wide, egg-shaped, broadest above the middle, briefly pointed at the tip, the margins toothless, basally tapered to the short leafstalk, with numerous lateral veins arising featherlike from the midvein. FLOWERS: solitary, short-stalked, arising from twigs of the previous year, green at first, turning brownish maroon; sepals 3, free of each other; petals 6, in two whorls of 3; stamens numerous, forming a spire around the few, separate female organs. FRUITS: large solitary or paired, fleshy, aromatic berries 2–6 inches long, 1–2 inches thick, green, spotted brown when ripe. SEEDS: one to several, flattened, brown, to 1 inch long.

DISTRIBUTION: Understory tree in low, sandy, or alluvial deciduous forests, flowering as early as March to May, the fruits ripening from July farther south to frost farther north. Dry summers like that of 1986 produce few pawpaws, at least around the District of Columbia. NE-NY-TX-FL. Zones 4–8.

UTILITY: The aromatic fruits are so tasty that it may be difficult to beat the possums and raccoons to them in autumn. To me, it ranks with the mayapple and passionfruit as good, edible fruits available in late summer, tasty, but different in flavor, and hardly comparable to any of our more conventional fruits. Like many Amerindian food plants, this one has no folk medicinal claims, at least as reported by Moerman. Some people are said to be allergic to pawpaws. Seeds and bark of the pawpaw and related species tend to have pesticidal activity. Nonetheless, I am told that the seeds are consumed by wild turkeys. Someone lost in the woods and plagued by head lice might apply crushed pawpaw seeds to the head. The seeds of this and many related species contain potent pediculicides. Caterpillars of the zebra swallowtail butterfly feed exclusively on young foliage of pawpaw trees, perhaps sequestering some of the toxic pesticides from the malodorous leaves. The bark contains similar natural pesticides called acetogenins which appear to act synergistically. Organic gardeners well endowed with pawpaw might try the leaves as insect and fungal repellant mulches. One USDA scientist warned me that two of his workers suffered reversible ophthalmic problems when working with these acetogenins. USDA Economist Neal Peterson tells me that there are places called Pawpaw in IL, IN, KS, MI, OK, and WV,

Asimina triloba (L.) Dunal

which states are locally still well endowed with pawpaw. I have heard from many reliable sources that the pawpaw is a good indicator of good ginseng hunting. I have yet, though, to find a ginseng near the pawpaws I have found wild (although I have cultivated root sports of pawpaw over my cultivated ginseng in Phenology Valley). Ginseng is not exactly a roadside weed. Pawpaw may be a weed with ''weedicidal'' potential.

AVENA FATUA L. (POACEAE) --- Wild Oat

DESCRIPTION: Annual grasses with fibrous roots, the stems grayish green, hairless, 10–40 inches tall. LEAVES: basal and then one at the node, alternating or spirally arranged along the stem, the blades 3–16 inches long, less than 1 inch wide, sometimes rough to the touch (scabrous), with no obvious leafstalk and no teeth; veins parallel. FLOWERS: pendulous in loose, spreading, terminal clusters (panicles) 6–16 inches long, with short bristles extending beyond the individual florets. FRUITS: hairy grains 1/4–1/3 inch long in ribbed green enclosures terminating in bristles as much as 1–1.5 inches long.

DISTRIBUTION: Rather ubiquitous where cereals (e.g., barley, oats, wheat) are grown in the northern tier, this weedy grass occurs in cereal fields, feedlots, and poorly maintained pastures, roadsides, and waste places, in all 48 states. Flowering mostly July to August, in Maryland at least. In western Canada, plants start flowering in early July, shedding seed by mid-August, the spread of flowering time exceeding that of the cultivated oat. WA-ME-CA-MD. Zones 3–7.

UTILITY: Often competing rigorously with other cereals, and having co-evolved with cultivated oats, wild oats can be used in many of the same ways as the cultivated oats. Wild oats are about 90% as nutritious as the "tame oat", but they may have more than 50% more fiber, for which oat bran has been so lavishly praised lately. Several studies indicate that oat bran has significant cholesterol-lowering benefits for those with seriously elevated serum cholesterol, but it takes 3–4 ounces of oat bran a day to provide the necessary beta-glucan (SN 137:330.1990). A gruel of wild oat, pumpkin, evening primrose, and groundnut seeds would have several chemopreventive compounds therein. American Indians have long consumed the fiber-rich wild oats. Wild oats are even sold as "mill oats" in the U.S. As a weed, wild oats can cut spring wheat yields by 30% in the Northern Plains states. Ironically, the herbicides used to control wild oats may also injure some wheat varieties (Hays, 1991). When I was a kid, I was often offered hot oatmeal as something to stick to my ribs. Maybe mother should have tried Grieve's Gruel (1974), boiling 1 ounce "tame" oatmeal in 3 pints of water, until it boils down to a quart. Strain, adding sugar, lemons, wine, and/or raisins (Grieve also recommended this as a "demulcent enema"). For a while, there was big press about oats helping one break the cigarette habit. This rumor has subsided. Seeds of wild or "tame" oats can be scorched as herbal coffee substitutes, devoid of caffeine or ground up and made into beer. In California, some 40,000 acres of wild oats are harvested for hay. It can also be used for pasturage. Reported hay yields might attain 2–3 tons/acre. Though oats are often grown near barley and wheat in the Holy Land today, oats do not seem

Avena fatua L.

to have been mentioned in the Bible. ''Tame'' oats are regarded as demulcent, diuretic, nervine, stimulant, and tonic, while wild oats have been used for female ailments. CAUTION: The needlelike tips of the chaff should be removed before grazers indulge. They are bad about lodging in the throat. WSSA lists 5 species of *Avena* as weeds.

BARBAREA SPP. (BRASSICACEAE) --- Early Wintercress (*B. verna*) and Yellow Rocket (*B. vulgaris*)

DESCRIPTION: Small winter annual, biennial, or rarely perennial herb, overwintering in a winter rosette (whorl of leaves arising at the ground level, with no obvious stem above ground, just the leafstalks), bolting to as much as 2.5 feet. LEAVES: of the rosette dandelionlike, cut to the midrib, making several lobes (*B. verna* with 5–10 pairs of lateral lobes, *B vulgaris* with 1–4 pairs, in addition to the terminal lobe), with an obvious leafstalk; stem leaves one at the node, alternating on the stem, reduced up the stem, becoming smaller, with fewer teeth and shorter leafstalks, clasping the stem. FLOWERS: in clusters at the tip of the stem and its main branches, if any; sepals 4; petals 4, independent of each other, yellow; stamens 6; ovary 2-celled. FRUIT: a slender, 4-angled or rounded green pod, 0.5–1.5 inches long (in *B. vulgaris*) to 1.5–3.0 inches (in *B. verna*), with a single row of seeds in each of the two cells.

DISTRIBUTION: Common in waste places, old homesites, railroad rights-of-way, roadsides, and last year's garden, flowering March to May, fruiting shortly after flowering, less desirable as a potherb after bolting. WA-ME-OR-SC. Zones 4–8.

UTILITY: A proverbial spring tonic, this is among the first edible greens to appear in weedy areas around Herbal Vineyard. The whole plant can be used for food, but older portions are often too bitter and stringy to be palatable. Older portions should be parboiled once or twice, perhaps with bicarbonate to tenderize. Fresh leaves during winter (hence the name wintercress) can be a welcome addition to a bland winter diet. Flower clusters can be cooked like broccoli. Seeds could be used as oil sources or sprouted. Young leaves and undeveloped shoots can be used, as is, in salads. Brill (n.d.) gives an interesting recipe for Wintercress-Stuffed Potatoes, mixing the cress with diced onion, garlic, dill, celery, paprika, rosemary, tarragon, and yogurt. Like other members of the BRASSICACEAE, these, as other cresses, contain several cancer-preventive compounds. Cherokee ate boiled "creasy" greens to purify the blood. Mohegans and Shinnecock took the leaf infusion every half hour for coughs (Moerman, 1986). Crushed leaves are poulticed onto bee stings and other skin disorders (Brill, n.d.). The plant can intoxicate horses. It is reputedly antiscorbutic, apertif, balsamic, depurative, diuretic, and vulnerary (Duke and Wain, 1981). CAUTION: Author, forager, lecturer, and spinner of tall tales, Doug Elliott gently criticized Steve Foster and me (1990) for our cautionary warning in the *Field Guide to Medicinal Plants:* although this plant is described as edible, studies indicating the possibility the herb may

Barbarea vulgaris R. Br.

cause kidney malfunctions dictate moderation or abstinence. I rather feel that such studies should be pointed out. New studies tomorrow might indicate major changes in my recommendations. I will continue, as I imagine Doug Elliott will, to enjoy my wintercress in winter.

BETULA SPP. (BETULACEAE) --- Birches

DESCRIPTION: Deciduous, often aromatic trees, the bark usually distinct, white, but not peeling off in *Betula alba,* flaky and peeling off in *B. nigra,* aromatic and peeling off in *B. lenta,* and yellow and peeling off in *B. lutea.* LEAVES: one at the node, alternating along the zigzag stems, 1.5–5 inches long, 1–2 inches wide, egg-shaped or nearly triangular, broadest below the middle, with jagged, sawlike teeth on the edges, basally rounded to or notched at the short leafstalk, the midrib with numerous lateral veins arching out toward the teeth. FLOWERS: in greenish, conelike structures, some male with 2 stamens, some female, the ovary 2-celled, with 2 terminal processes. FRUITS: of winged nutlets borne in the brownish, hoplike female cones.

DISTRIBUTION: Some species occur regularly in alluvial forests, others in deciduous hardwood or even coniferous forests. Flowering March to May, before or with the leafing out, the seeds ripening in the cones in summer. *Betula nigra* (River Birch or Black Birch): IA-CT-TX-FL. Zones 4–9

UTILITY: Easily recognized in winter, the various birches can be drawn upon at all seasons for one thing or another. In early spring, the trunks can be tapped for sugar water like maples. Early Americans often described the Indians making a hole in the trunk and sucking out water whenever they were thirsty. In summer I have had little luck. The water extracted from the trunks can be boiled down to make sugar water, and then fermented to make birch booze or vinegar, depending upon which yeasts move in first. Inner bark of this, like that of many other trees, and all birches (Naegele, 1980) can be ground up and made into flour for making bark bread. Never remove a complete circle of bark for this purpose, because it can kill the tree. Better peel off the outer bark to use for tinder, then gouge out the succulent inner bark for foodstuff, taking vertical strips. Leaves, twigs, and buds of the aromatic birches make good teas, similar to wintergreen tea. Cherokee took river birch tea for colds, dysentery, stomach distress, and an ailment described as "milky urine". Chippewa used the bark tea for stomachache. (Moerman, 1986). Alabama Indians used the bark tea to treat horse hooves (Duke, 1986). The cherry birch, *B. lenta,* loaded with the aromatic (and poisonous) methyl salicylate, was used by the Osage Indians for breast complaints, colds, coughs, scrofula, and sores. But like other salicylates, this too can cause stomach distress. Potawatomi preferred sapling yellow birch as poles for their wig-wams. Buds and seed are eaten by grouse, seeds by chickadees, finches, sparrows, and other songbirds, and by chipmunks, squirrels, and wood rats.

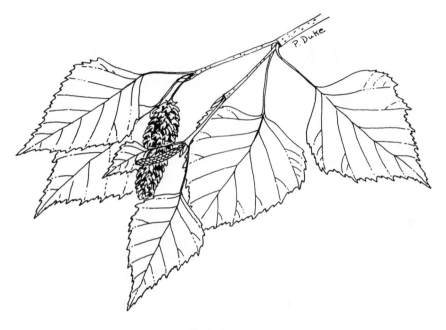

Betula sp.

Deer, elk, and moose browse the foliage and twigs; beavers, porcupines, and rabbits the bark or wood. I wouldn't hesitate to sample any of these (plant or animal) in an emergency. WSSA lists additionally the Gray Birch, *B. populifolia* Marsh., as a weed.

BRASSICA NIGRA (L.) W.J.D. Koch   (BRASSICACEAE)
--- <u>Black Mustard</u>, Wild Mustard

DESCRIPTION: Low annual, biennial, or perennial herbs, often with a winter rosette and taproot. LEAVES: of the rosette dandelionlike, cut several times to the midrib, leaving several stalkless lobes; stem leaves stalked or stalkless, not as deeply lobed as the rosette leaves, with rounded lobes or teeth. FLOWERS: in long clusters at the tip of the stem and main branches, if any; sepals 4; petals 4, yellow, forming the cross characteristic of the family Cruciferae (= BRASSICACEAE); stamens 6; ovary 2-celled. FRUIT: a long-stalked pod, round or 4-angled in cross section, with 1 row of round to oblong seeds in each of the 2 cells of the ovary.

DISTRIBUTION: All the American species are introduced weeds, frequent in gardens, old fields, meadows, waste places, and along roads and railroads, flowering from March to May, fruiting April to June, some species with a second flowering period in autumn. *Brassica nigra:* WA-ME-CA-FL. Zones 3–8.

UTILITY: Another of our spring tonics, waking up the doldrums of the slowed peristalsis of winter, the wild mustards are among the most copious wild greens around Maryland and probably farther north. All parts of the plants may be eaten, depending on your tastes. I have eaten all the various parts; the petals are less peppery (and flavorful). Green pods sometimes taste good raw, but they can be quite tough and fibrous when mature (Fat America needs the fiber and the crucifer). I have cooked the young flower clusters like broccoli and added them to soups and salads. Bitter batches are improved by boiling with several changes of water. A country boy, I like the pot liquor, especially with cornbread, better than the potherb. Seeds may be ground up as pungent wild mustards, or may be sprouted. Broccoli, brussels sprouts, cabbage, cauliflower, collards, kale, kohlrabi, mustard, and turnip greens all belong to the genus *Brassica*, very important in anticancer diets, according to modern reports from the NCI. Ironically, the anticancer compounds, known as isothiocyanates, are also reported to cause cancer under other circumstances. As Paul Stitt was publishing his *WHY GEORGE SHOULD EAT HIS BROCCOLI,* in which he enumerated 33 cancer-preventive compounds in broccoli (a variety of *Brassica oleracea*), world-famed Bruce Ames was enumerating almost as many carcinogenic, clastogenic, and/or mutagenic compounds in cabbage (another variety of *Brassica oleracea*). Still, the high-fiber-, beta-carotene-, ascorbic acid-, sulfur-containing compounds and tocopherols can all contribute to the cancer-preventing activity, leading to one of my popular articles, ''COLE'S ROLE'' (Duke and Barnett, 1989). Remember: all things in moderation, even cancer-preventive diets! Other

Brassica nigra (L.) W. J. D. Koch

Brassica species on the WSSA list are *B. hirta* (White Mustard), *B. juncea* (Indian Mustard), *B. kaber* (Wild Mustard), *B. rapa* (Birdsrape Mustard), and *B. tournefortii* (African Mustard). Most of these can be used interchangeably, by the forager at least. Hybrid *Brassica* are being used to smother weeds, e.g., in corn, dying back in 5 weeks, having smothered 80% of corn weeds (*Science News,* March 16, 1991).

BROMUS JAPONICUS Thunb.   (POACEAE) ---
Bromegrass, Cheat, Cheatgrass, Japanese Bromegrass

DESCRIPTION: Erect or spreading winter annual grasses with fibrous roots, the stems grayish green, hairless, 12–40 inches tall. LEAVES: basal and then one at the node, alternating or spirally arranged along the stem, the blades 5–20 inches long, less than 1 inch wide, slightly hairy and sometimes rough to the touch (scabrous), with no obvious leafstalk and no teeth; veins parallel. FLOWERS: numerous, in loose, spreading, terminal clusters (panicles) 2–8 inches long, with short bristles extending beyond the individual florets. FRUITS: hairy grains 1/5–1/4 inch long in ribbed green enclosures terminating in bristles 1/5–1/2 inch long.

DISTRIBUTION: Weedy species, introduced from Europe, now widespread in the U.S., in cereals (e.g., barley, oats, wheat), meadows, pastures, roadsides, and waste places, with one or another species in all 48 states. Not so serious weeds as *Agropyron* and *Avena*. Flowering mostly May to August, depending on the species. *Bromus commutatus* ("Upright-" or "Hairy Chess") flowers June to July, *B. secalinus* ("Cheat") June to September in Maryland; *B. japonicus* ("Japanese Brome grass or chess") June to August: WA-ME-CA-FL. Zones 3–8.

UTILITY: Often competing with other grasses and cereals, bromegrasses and chesses can be used in many of the same ways as cultivated cereals. Several studies indicate that oat bran has significant cholesterol-lowering benefits for those with seriously elevated serum cholesterol. I suspect that the chaff and bran in wild grasses can also help lower cholesterol. American Indians have long consumed the fiber-rich *Bromus* seeds. Seeds of the California Brome, *Bromus carinatus,* and *B. rigidus* were eaten in California, while seeds of *B. carinatus* were used for pinole in California, Nevada, and Utah. California Brome has been suggested (Facciola, 1990) as a potential perennial grain crop able to survive summer droughts in California, a vital point as California begins water rationing in early 1991. The biennial Chilean species, *B. mango,* confusingly called "mango", was thought to be extinct, but has been rediscovered. Araucano Indians made an unleavened bread from ground seeds of the "mango" and also a beverage called chicha (Facciola, 1990). Seemingly, the seeds of any *Bromus* species can be scorched as herbal coffee substitutes, devoid of caffeine, or ground up and made into beer. Brome grasses are occasionally used for hay in Oregon and Washington, but are more often viewed as foe than friend. They can also be used for pasturage.

Bromus japonicus Thunb.

Reported hay yields of *Bromus inermis* have exceeded 2.5 tons/acre. Amerindian folk usages are few, *B. ciliatus* used by the Iroquois, *B. tectorum* used as a face-wash (by ''God-Impersonators'') (Moerman, 1986). WSSA lists nearly 20 species as weeds.

CANNABIS SATIVA L.   (CANNABINACEAE) ---
Cannabis, Grass, Marijuana

DESCRIPTION: Tall, bushy, sticky, aromatic, annual weeds to 15 feet tall, with a much-branched taproot. LEAVES: two opposing each other low on the stem, one at the nodes above, palmately compound, i.e., with 3–9 hairy saw-toothed, willowlike leaflets arising from a leafstalk that is about as long as the leaflets, 2–6 inches long. FLOWERS: male flowers, several in clusters towards the tips and coming from the axils of the upper reduced leaves, nearly as long as the leaves with which they are associated; female flowers fewer, in smaller clusters on entirely different plants, with hairy green structures with 2 elongate processes towards the tips. FRUITS: small, greenish, saclike structures, each containing a single lentil-shaped seed.

DISTRIBUTION: Fertile, moist soils as in farmyards, fencerows, old fields, roadsides, and streamsides, occurring as a waif or clandestine cultivar, beginning to flower in July in Maryland, flowering and fruiting until frost or harvest. WA-MA-CA-FL. Zones 4–10.

UTILITY: While we in the U.S. tend to think of this as a drug plant, each plant valued at $3000 for drug-enforcement calculations, it is elsewhere cultivated as an oilseed, fiber, and medicinal species. Frequent in birdseed mixes, the seeds can be toasted and eaten. In Thailand and Laos, I was served soups in which *Cannabis* tops were one of several "vegetables", and I have been served cookies containing *Cannabis* here in the U.S. In Japan, seeds, called "asanomi", are utilized in deep-fried tofu burgers called "ganmo". Seemingly, the seeds of *Cannabis,* like those of the opium poppy, *Papaver somniferum,* are not narcotic. Breaking scientific, Naegele (1980) gives dosage rates for "The Weed", for eating, 300–480 mg/kg body weight; for smoking, 200–250 mg/kg. Since it is illegal, you'd best eat another weed. As a jazz musician back in college, I sometimes indulged in the "weed" as a fumitory, though a bit alarmed by the verse:

> There was a young man smoked pot
> Until his hormones got shot;
> His testes undercharged,
> While his breasts, they enlarged;
> Was he a young man or not?

Cannabis, or its active ingredient (tetrahydrocannabinol, or THC), is still useful in glaucoma and as an antinauseant in chemotherapy. It is such a productive weed that even today, some Americans advocate its growth for biomass. There might be merit in widespread planting of the low-THC weedy strains in areas where high-THC marijuana is illicitly cultivated, as I proposed back in 1970. The pollen from the low-THC plants might genetically dilute

Cannabis sativa L.

the THC-strength in seeds of the high-THC clandestine females. This presumes that resulting hybrids would have a lower THC content than the high-THC parent. Gradually, volunteer plant populations would be diluted as regards their THC content. They say that George Washington introduced fiber hemp along the Potomac, and people are still harvesting it to this day. Watch out for booby traps when harvesting this most dangerous weed, California's No. 1 Crop.

CAPSELLA BURSA-PASTORIS (L.) Medic.
(BRASSICACEAE) --- Shepherd's-purse

DESCRIPTION: Low, inconspicuous winter annual, with a basal winter rosette (a circular cluster of "ground-hugging" leaves arising from a central point under which lies the root and often the crown of the plant). LEAVES: of the rosette 2–5 inches long, toothed or pinnately lobed (cut to the midrib, leaving a featherlike display of lobes), relatively long-stalked, the stem leaves smaller, with proportionately smaller stalks, and shallower lobes; uppermost leaves with their bases nearly surrounding the stem, sometimes with star-shaped (stellate) hairs. FLOWERS: small, white, in elongate clusters to 12 inches long terminating the main stem and often branches as well; flower stalks to nearly an inch long; sepals 4; petals 4; stamens 6; ovary 2-celled, with around 6 small ovules in each cell. FRUITS: triangular or notched at the tip to give a heart-shaped effect, green, long-stalked, with about 6 small seeds in each flattened half of the triangle.

DISTRIBUTION: A common weed around houses, old fields, pastures, gardens, and the like, one of the many spring members of the mustard family (BRASSICACEAE) available to foragers. Commonly, it flowers in early spring, dying back in the heat of summer. New seedlings emerge in fall, some of them sometimes flowering on warmer winter days. WA-ME-CA-FL. Zones 4–8.

UTILITY: Like many other members of the mustard family, this offers a spicy potherb, which may double as a spring tonic, perhaps with some cancer-preventive activity as well. Tender plant parts are pleasantly peppery when eaten raw (also cancer preventive according to the NCI); older plants are tough enough to warrant a pinch of bicarbonate of soda to tenderize the fibers. Indians ground the minute seeds to make a nutritious flour. Had I gone to the trouble to gather the seeds, I think I would as soon use them for sprouts, perhaps getting proportionately more vitamin, fiber, and protein and less fat that way. Storage fats in seeds are often mobilized in germinating. Japanese are big on sprouts of the radish, also in the mustard family (which also includes cabbage, cauliflower, broccoli etc). Japanese consider shepherd's purse an essential condiment in their barley-rice gruel, a ceremonial dish on January 7 (Facciola, 1990). Several foraging books suggest the use of the fruits (and/ or seeds) as a pepper substitute for addition to soups and stews. Facciola (1990) comments that the fresh or dried roots are used as a ginger substitute. Reducing the plant to ashes by burning in an enclosed container yields a gray substitute for salt. In ashing, almost everything combustible is driven off,

Capsella bursa-pastoris (L.) Medic.

leaving the minerals in the ash, including sodium, potassium, etc. Such ashes might also function as a substitute for bicarbonate in tenderizing fibrous plants. In a camping situation, I don't hesitate to add ashes from my campfire (if I have stuck to the better firewoods like oak, hickory, maple, beech, etc.) to a stew for salt and tenderizer.

CARDAMINE BULBOSA (Schreber) BSP.
(BRASSICACEAE) --- Bittercress, Bulbous Cress, Flag

DESCRIPTION: Small, upright perennials 6–20 inches tall, from a cluster of small bulbs. LEAVES: of the rosette long-stalked, not toothed or lobed, egg-shaped to essentially round, 0.5–2 inches long, grading into 4–14 stem leaves, the stem leaves proportionately longer, and irregularly toothed or lobed, with shorter stalks or no stalks at all. FLOWERS: few, in lax, long clusters at the tip of the stem and occasionally main branches as well; sepals 4; petals 4, white, much longer than the sepals; stamens 6; ovary 2-celled. FRUIT: an erect, long, narrow pod to ca. 1 inch long, round in cross section or nearly so, stalked, with 1 row of flat seeds in each of the 2 cells.

DISTRIBUTION: Occasional in marshes, slow, small streams, and rich, moist woodlands, often in dense shade, flowering and fruiting April to June and dying back before frost, at least near me in Maryland. ND-MA-TX-FL. Zones 3–8.

UTILITY: Once you locate this woodland species, you can harvest it judiciously at any stage. I like the roots as a pungent nibble, and the vegetative parts as a salad or potherb. Roots, even the pods, can be ground as a horseradish substitute, improved by adding vinegar if you have any in your survival kit. I will guarantee that the fresh roots are tear-jerkers, like horseradish. Not wishing to make a whole meal of this, I do like to spice up a dull slice of venison or jerky, perhaps even pemmican, with Father Nature's horseradish sauce. Though somewhat peppery, the pods, and especially the leaves, are bland, compared to the roots. Facciola (1990) notes that the young leaves, shoots, and flower buds of the cuckoo bittercress have a pungent cresslike flavor and can be added to fresh green salads, sandwich spreads, sauces, or served alone au natural or with oil and vinegar. Various *Cardamine* species were used by Amerindians in their medicine. Cherokee used them for colds, headaches, and sore throats. Iroquois used them for cardiac palpitations, chest pain, dyspepsia, fever, gas, headaches, and poor appetite (Moerman, 1986). Europeans, regarding the cuckoo bittercress as antiscorbutic, depurative, diaphoretic, diuretic, nervine, and stimulant, used it for cramps, scurvy, and tumors. The vitamin C would explain its use for scurvy, and its many cancer-preventive compounds, characteristic of the mustard family, may explain its anticancer reputation. The weedy species listed by the WSSA, *C. hirsuta* (Hairy Bittercress), *C. parviflora* (Smallflowered Bittercress), *C. pensylvanica* (Pennsylvania Bittercress), and *C. pratensis* (Cuckoo Bittercress), are probably as edible as the bulbous bittercress, but lack the succulent, pungent

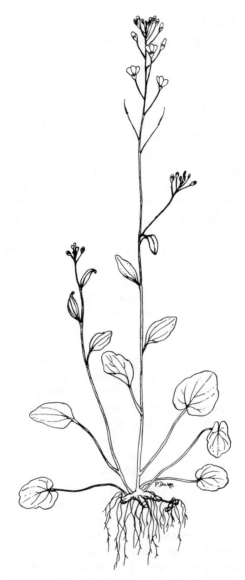

Cardamine bulbosa (Schreber) BSP.

roots. CAUTION: While I have eaten any species I encountered, Moerman (1986) relates that the Iroquois used the root of *C. bulbosa* for poisonings, the plant of *C. concatenata* to mesmerize, the root of *C. douglasii* in witch-craft.

CARPINUS CAROLINIANA Walter (BETULACEAE) --- American Hornbeam, Blue Beech, Ironwood

DESCRIPTION: Small, deciduous understory trees to as much as 40 feet tall, the bark smooth, sinuous, and bluish gray (hence the name blue beech). LEAVES: one at the node, alternating on the zigzag stem, narrowly egg-shaped, broadest below the middle, 2–5 inches long, 1–2 inches wide, sharply pointed at the tip, basally rounded to the short leafstalk, with sawlike teeth along the margin, with several lateral veins extending featherlike from the midvein. FLOWERS: inconspicuous, appearing before the leaves, the male and female flowers in lax, conelike clusters, the male on short spur branches, with 3–12 forked stamens, the female paired toward the end of branches; ovary 2-celled with 2 terminal processes. FRUIT: small nutlets enveloped in greenish, leafy appendages, later turning brown and functioning winglike for dispersal.

DISTRIBUTION: Frequent in deciduous forests, especially along streams in the piedmont and mountains, flowering March to May, the seeds ripening from late August to frost. MN-ME-TX-FL. Zones 4–9.

UTILITY: Lacking in most foraging books, the blue beech is easy to recognize, hard to confuse with anything poisonous, and readily available from August to December or later. With its smooth blue-gray trunk, it is rather suggestive of the true beech (*Fagus grandifolia*), beechnuts of which are more edible and less weedy. The trunk of the blue beech is not even and smooth like that of the true beech, being marked with musclelike ridges often termed "sinewy". The leafy structures come off with the small nutlet. I gather a handful, roll them in my hand, gradually crushing the leafy envelope and liberating the small nutlet. It's not much of a nut, but man shall not live by greens alone. The seeds are also eaten by ducks, grouse, pheasant, quail, and warblers. Deer browse the twigs and foliage; beavers, rabbits, and squirrels use the bark, seed, and/or wood. Even some of the medicinal uses of *Carpinus* are somewhat vague. Moerman (1986) reports its use for "cloudy urine" and "navel yellowness". The astringent inner bark was used to staunch bleeding. Delaware Indians used the root or bark infusion for general debility and female ailments. Iroquois used it for childbirth, and used the bark chips in a polyherbal formula for tuberculosis. Iroquois also used it for "big injuries" and "Italian itch" (Moerman, 1986). The light brown wood, difficult to chop and work, is used for tool handles and other small articles as well as firewood. Blue beech is not listed as a weed by the WSSA, but it is a weed in my ginseng patch.

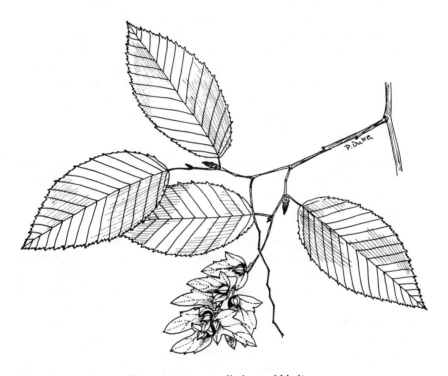

Carpinus caroliniana Walter

CARYA SPP. (JUGLANDACEAE) --- Hickory

DESCRIPTION: Tall deciduous trees, the bark often close and gray or flaking off in large strips, giving the trunk a conspicuous shaggy appearance. LEAVES: feather-compound (pinnate), one at the node, alternating along the thick twigs, with 3–11(–21) short-stalked, egg-shaped leaflets, in pairs along the midrib of the compound leaf, broadest at, below, or above the middle, pointed (acute) at the tip, rounded where they join the midrib (rhachis) of the compound leaf, with many teeth along the edges, and numerous lateral veins arising featherlike from the midrib of individual leaflets. FLOWERS: inconspicuous, in conelike clusters, the male flowers near where new growth of the current year commenced, with 3–8 stamens, the females in clusters of 2–10, towards the tips of new twigs, each ovary with 2 terminal processes. FRUIT: a very hard nut enclosed in a greenish to brownish woody husk that tardily splits into 4 quarters, starting at the tip.

DISTRIBUTION: Common, often dominant, tree in the Eastern Deciduous Forests, often with oaks, less often with beeches and maples, flowering in April and May, the fruits not maturing until near frost. *Carya ovata:* MN-NY-TX-AL. Zones 3–8.

UTILITY: Squirrels often compete with foragers for these valuable nuts, once a staple winter food with some Indian tribes, such as the Dakota, Omaha, Pawnee, and Winnebago. Good foragers can find the nuts all winter, raiding the caches of other animals. It takes a lot of energy to get to the energy-rich seed, high in fat content, but with mostly unsaturated fatty acids. Indians sometimes shortcut the husking process and smashed the whole, unhusked nut, boiling the macerate, and skimming off the oil that floated to the top of the pot. One soon learns that those whose husks split nearly to the base in nature often have the sweetest meats. Squirrels, knowing this too, apparently squirrel these away first, leaving the unopened, bitterer nuts overwinter on the forest floor. I'll wager that the nuts squirrelled away will average less bitter than those remaining on the forest floor for the forager. Some hickories, sources of excellent firewood and charcoal, were also tapped for sugar, like the maples. Indians even extracted a sweetener by boiling hickory wood chips, which could still be used for fuel or charcoal. Indians were said to have eaten "tendrils of young roots". Fox Indians used bark tea of *C. cordiformis* for simple maladies, bowel and urinary problems. Comanche used leaves of the pecan, also a *Carya, C. illinoensis,* for ringworm; Kiowa took the bark decoction for tuberculosis. Cherokee used *C. laciniosa, C. pallida,* and *C. tomentosa* for colds, female obstructions, and to relieve the pains of polio.

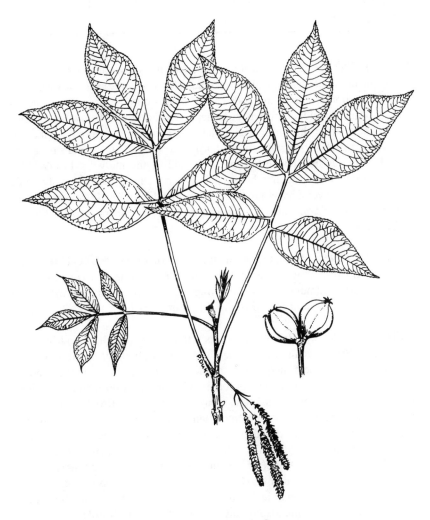

Carya glabra (Mill.) Sweet

Considering the hickory choleretic, diaphoretic, emetic, and stomachic, they chewed the astringent bark for sore mouths and applied the bark to cuts. Delaware used *C. ovata* for female ailments and general debility, Iroquois for arthritis and worms. Hickory is one of the best firewoods, and famous for smoking barbecues. WSSA lists as a weed only *Carya glabra* (Pignut Hickory).

CASTANEA SPP.   (FAGACEAE) --- Chestnut, Chinquapin

DESCRIPTION: Deciduous shrubs or small trees (the latter largely wiped out by the chestnut blight, but being replaced by resistant species or varieties). LEAVES: simple, one to the node, alternating along the stem, 2–10 inches long, 1–4 inches wide, narrowly egg-shaped, broadest at or below the middle, hairless beneath in chestnuts, slightly hairy in chinquapins, pointed at the tip, tapered to the short leafstalk, with sharp teeth on the edges, and prominent lateral veins, arising featherlike on each side of the midvein. FLOWERS: female in conelike structures below the elongate cones (catkins) of the male flowers, both developing after the leaves. FRUITS: Female flowers developing into a burr containing 1 (chinquapin) or 2–3 (chestnut) shiny, brown, edible nuts.

DISTRIBUTION: Sprouting from the roots of diseased parent trees, the chestnut still resides in our mountains, and the chinquapin is more characteristic of dry piedmont forests, flowering June to July, fruits maturing up to frost. *Castanea dentata:* MN-ME-AR-FL. Zones 3–7.

UTILITY: Old Mr. Brooks across the street from my childhood home in East Lake, Birmingham, was a lonely old man of about 80, I guess, when I was 5. He knew where there were still plenty of chestnuts up on the ''mountain'' across the tracks. It was probably Mr. Brooks who instilled the love of the forest, the ''mountain'', and foraging in me. Mr. Brooks and most of the chestnuts are gone, but the ''mountains'' remain forever as their monuments. Thanks! Nuts of the alien and native species of chestnuts and chinquapins are doughy, but edible after roasting. My Chinese chestnuts are prolific producers, but 99 + % of the nuts are consumed by small insect larvae, as evidenced by their exit holes. Chestnuts in general are less tasty and higher in carbohydrates and lower in oils than our hickories and walnuts. Still, early settlers roasted the chinquapins and used them as a chocolate substitute. Indians used chestnuts and chinquapins in breadstuffs and soups, the Iroquois making a coffeelike beverage from the roasted nuts. Early settlers valued the chestnut for timber, hoops, posts, and staves, and the bark for tanning and dying leather. Indians used it to tan deer skins. The black sap of older trees was used for ink (Erichsen-Brown, 1979). Cherokee used the tannin-rich chestnut leaves in some of their cough syrups and applied them to cuts and wounds. They used infusions, often of the bark and/or leaves, for heart and stomach distress and for bleeding following childbirth. The Cherokee had different uses for the chinquapin, using the leaves for chills, cold sweats,

Castanea mollisima Blume

fever, fever blisters, and headache. Southerners used the chinquapin for ague. Iroquois used chestnut wood powder for baby powder for chafed babies and added the bark to dog food as a dewormer. Mohegans used the leaf infusion for colds, rheumatism, and whooping cough. No *Castanea* are listed by WSSA, though both can be weed trees from root sprouts.

CENCHRUS SPP. L.   (POACEAE) --- <u>Sandbur</u>, Sandspur

DESCRIPTION: Erect or spreading winter annual grasses with fibrous roots, the stems grayish green, hairless, 12–40 inches tall, sometimes rooting at the lower nodes. LEAVES: basal and then one at the node, alternating or spirally arranged along the stem, the blades 1–7 inches long, much less than 1 inch wide, rarely hairy at the junction of leaf and stem, sometimes rough to the touch (scabrous), with no obvious leafstalk and no teeth; veins parallel. FLOWERS: few, scattered, in loose ascending or spreading terminal 2–4-flowered clusters, the inflorescence 1–4 inches long. FRUITS: included within spiny, nearly globose burrs ca. 1/3–1/4 inch long, containing 1–3 seeds, the spines ca. 5–40 per burr.

DISTRIBUTION: Weedy species of open, sandy soils and waste places, especially near wool factories. A very pesky weed because the spines stick to the socks or penetrate the bare skin, even the tough callus underfoot. Flowering mostly June to October in the Carolinas, July to October in Maryland, and occurring mostly as beach bums or wasteland waifs further north. *C. pauciflorus:* CA-MD-AZ-FL. Zones 5–10.

UTILITY: Many years ago, as a beach bum, botanizing in the coastal plain of the Carolinas, I learned that one could make the picking of the burrs from the socks more pleasant by eating the 1–3 kernels that occur in the burr, which can be cracked with some practice. Several studies indicate that oat bran has significant cholesterol-lowering benefits for those with seriously elevated serum cholesterol. I suspect that the chaff and bran from wild grasses, even sandspurs, can also help lower cholesterol. North American Indians are not reported to consume the seed (Yanovsky, 1936), or use the plant medicinally (Duke, 1986; Foster and Duke, 1990; Moerman, 1986). Duke and Wain (1981) report medicinal uses for other species elsewhere, as an anodyne, diuretic, emollient, and vulnerary. Mexicans use the dune sandspur for tumors. Kalahari Africans reportedly ground the seed of an African species to make porridge. Seeds can probably be scorched to make herbal coffee, devoid of caffeine, or ground up and made into beer or breadstuffs, though the collecting and processing would be tedious. Before the burrs harden, the weed makes reasonable fodder as well. The <u>Buffelgrass</u>, which looks more like a foxtail or millet than a sandspur, has produced 10 tons/acre of biomass. The species listed by WSSA, *C. echinatus* (<u>Southern Sandbur</u>), *C. incertus* (<u>Field Sandbur</u>), *C. longispinus* (<u>Longspine Sandbur</u>), and *C. tribuloides* (<u>Dune Sandbur</u>), are

Cenchrus longispinus (Hack.) Fern.

probably equally obnoxious, as foods and weeds, but still better than nothing to a starving man or woman. CAUTION: Seeds have been reported to contain ergotlike alkaloids when fungally affected. Foragers should avoid all fungally contaminated wild foods.

CHENOPODIUM ALBUM L.   (CHENOPODIACEAE) ---
Common Lambsquarter

DESCRIPTION: Erect, annual herbs to 8 feet tall, the stems rather succulent, without hairs, but often with minute mealy protuberances which can
be rubbed off. LEAVES: simple, one at the node, alternating on the stem,
2–5 inches long, unevenly diamond-shaped, pointed at the tip, with irregular
teeth or small lobes on the edges, tapered to the leafstalk, which is usually
shorter than the blade; the undersurface often with a whitish or mealy cast
which can be rubbed off. FLOWERS: small, greenish, in elongate simple or
compound clusters toward the tip of the stem and major branches; sepals 3–
5; petals 0; stamens 5 or fewer; ovary simple, free of the sepals, with 2–3
processes at the tip. FRUIT: a small, papery, rounded pod, opening around
the tip, each containing a single shiny black seed.

DISTRIBUTION: Common weed, often in pure stands, in old fields, pastures, gardens, waste places, forest edges, in full sun or partial shade, starting
to flower in June, fruiting shortly thereafter, with another burst of flowering
and fruiting from a second "crop" in late summer, fruiting until frost. WA-
ME-CA-FL. Zones 4–8.

UTILITY: One of my most copious potherbs in spring, the lambsquarter
faithfully invades the garden and pasture each spring. Most amusing it is to
watch the wife pulling up the lambsquarter to favor the spinach. I'll wager
she couldn't tell them apart blindfolded if they were prepared in the same
manner, i.e., picked, washed, boiled lightly, and smothered in butter. If she
would just enjoy the lambsquarter, she would have ten times as much, still
enjoying it in summer, long after the related spinach has bolted. (Beets, chard,
lambsquarter, mangold, and spinach all belong to the CHENOPODIACEAE.)
Tops can be eaten before and after flowering. There are other nutrients in the
small seeds which can be used in porridge or ground up like any other cereal
to make a flour. Before the seeds are fully ripe, I gather them in the "milk"
stage, merely stripping off the clusters to add sparingly (they are a bit bitter,
like the raw leaves) to salads and soups. Pioneers added the seed to breads,
cookies, muffins, and pancakes. Mitich notes that they were in the gruel of
the last supper of Tollund man, before he was pitched into an Irish bog.
(WT2:550.1990) Facciola (1990) mentions that the sprouted seed are edible,
a suggestion he, like me, makes for most species of which both the seeds
and foliage are generally recognized as food (GRAF). Lambsquarter is said
to be useful in treating vitiligo. Cherokee ate the greens to "stay healthy"
(Moerman, 1986). Fox used the root infusion for urethral itch. Iroquois poulticed it onto burns. WSSA lists 18 weed species of *Chenopodium*, most of
which could be considered as food by the forager. CAUTION: The Mexi-

Chenopodium album L.

cantea, *C. ambrosioides,* might best be treated as medicine, since it contains poisonous compounds once used to treat worms. It is also added to beans to reduce the gas generated by their ingestion. Some lambsquarter and pigweed relatives can cause nitrite and nitrate poisoning in cattle who overgraze them. Mitchell and Rook (1979) mention a pellagralike syndrome in starving individuals who consumed *Chenopodium.*

CHRYSANTHEMUM LEUCANTHEMUM L.
(ASTERACEAE) --- <u>Oxeye Daisy</u>

DESCRIPTION: Low, perennial herbs 1–2.5 feet tall, often with a basal rosette from a short underground rootstock. LEAVES: of the rosette comblike (pectinate), 1.5–6 inches long; stem leaves one at the node, alternating on the stem, strap-shaped, hairless, broader above the middle, pointed at the tip, irregularly toothed or lobed on the edges, stalkless, the leaf base partially embracing the stem, with one midvein and obscure lateral veins branching off like plumes on a feather. FLOWERS: many, gathered into the tight cluster, which "daisy" the amateur might regard as a single flower; central flowers yellow with 5 sepals, 5 petals, 5 stamens, and a single ovary with 2 processes at the tip; marginal flowers with the big white strap, pulled off as we recite "she loves me, she loves me not." FRUIT: a minute, sunflowerlike seed with each of the central flowerlets.

DISTRIBUTION: Common weed or wildflower, depending on your disposition, in meadows, pastures, and waste places, flowering from April to July and sporadically up to frost. WA-ME-CA-FL. Zones 4–8.

UTILITY: I include the daisy, not because it is a superfood, but because it is superabundant and easy to recognize. Reading through the other foraging books, the kindest words might suggest that it makes an interesting addition to salads or that it makes a palatable salad itself. As one book aptly puts it, fondness for this "salad" must be an acquired taste (Fernald and Kinsey, 1958). Orientals seem to enjoy their species more than I enjoy mine. I have indulged in raw young leaves and early flower heads, but I can't recommend them. Orientals might. Italians eat the young leaves in salads (Grieve, 1974). Facciola mentions that "flowers" can be used like dandelions in home wine-making. Shinnecock Indians reportedly made a wine of daisy and dandelion flowers as a tonic (Moerman, 1986), apparently an acquired trait. Grieve (1974) mentions that European country people steeped the herb in ale to take for jaundice. Horses, sheep, and goats are said to eat the plant, cows and pigs to refuse it (Grieve, 1974). Menominee used the plant to treat fever. Mohegan used the tea of the whole plant or just the flowers as a spring tonic. Quileute used a decoction of the dried flowers cosmetically, as a wash for chafed hands. Europeans regard the daisy as astringent, diuretic, emmenagogue, and febrifuge, using it for asthma, cancer, catarrh, cramps, female ailments, fever, nightsweats, tuberculosis, whooping cough, and wounds (Duke and Wain, 1981; Grieve, 1974). CAUTION: The plant is said to cause dermatitis in sensitive people. I have had no problems. As with all flowers,

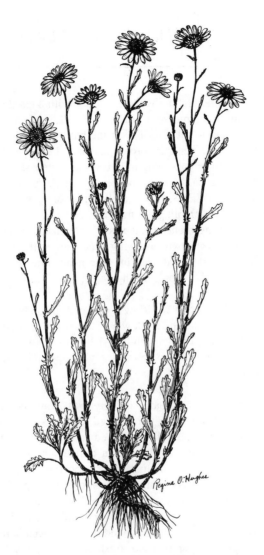

Chrysanthemum leucanthemum L.

especially members of the daisy or dandelion family, readers are again advised that people sensitized to pollen should exercise caution when ingesting flowers. Most male or bisexual flowers contain thousands of pollen grains. Even crude honey sometimes contains enough pollen to induce allergic reactions, even though honey is one folk ''remedy'' for allergy.

CICHORIUM INTYBUS L.   (ASTERACEAE) --- <u>Chicory</u>, Succory

DESCRIPTION: Perennial herbs to 5 feet tall, with a basal rosette, a deep taproot, a white milk oozing from breaks. LEAVES: of the rosette dande-lionlike (irregularly cut to the midrib near the base, the midrib often reddish and channeled, much longer than broad), to over 12 inches long, hairy on the underside of the midrib, exuding milk if broken; stem leaves smaller, one at the node, alternating on the stem, becoming smaller up the stem, with no obvious leafstalk. FLOWERS: blue, rarely white or pink, in dense, many-flowered heads, each resembling a single flower, solitary or paired towards the ends of the flowering branches; sepals 5; stamens 5; ovary solitary, em-braced by the sepals, with two diverging processes at the tip. FRUITS: of small, sunflowerlike nutlets crowded into the heads after flowering; nutlets brown or black, rounded, square, or angular in cross section.

DISTRIBUTION: Common wildflower or weed in sunny meadows, road-sides, and waste places, starting to flower in June, and flowering until frost. Flower heads not usually opening until 8–10 o'clock, often closing for the day at noon on sunny days, but sometimes staying open on cloudy days. WA-ME-CA-AL. Zones 4–8.

UTILITY: Known to the Romans, chicory was mentioned by such early historians as Horace, Ovid, Pliny, and Virgil, and was early consumed as a vegetable or salad (Grieve, 1974). Scorched roots, bitter as they are, have been used as a coffee, coffee adulterant, or coffee substitute, depending on your point of view. It is a shame that some good food can't be made from these easily available roots, abundant, easily harvested by pulling after a rain, and easy to recognize. One of my favorite wild vegetable broths (bouillons) is made of chicory flowers, red clover flowers, and wild garlic flower heads (unopened), all easily available in Maryland in June. Facciola reminds us that the attractive blue flowers can be used, fresh or pickled, in salads. The early spring shoots are almost as bitter as those of dandelion, relished by some, but too bitter for my jaundiced palate. Shoots, blanched by covering for weeks to keep light off, seem both more attractive and palatable, like the better and much more expensive Belgian endive so ridiculed by Vice President Quayle. Chicory Tea, 1 ounce of root to a pint of water, was a European folk remedy for gout, gravel, hepatomegaly, jaundice and rheumatic complaints (Grieve, 1974). Cherokee Indians use the root in nerve tonics and as a poultice on chancres and fever sores (Moerman, 1986). Duke and Wain (1981), citing chicory as alexiteric, aperient, cholagogue, depurative, diuretic, emmen-agogue, laxative, refrigerant, sedative, stomachic, sudorific, and tonic, men-tion its use for cancer, fever, gall problems, hepatosis, inflammations, nausea,

Cichorium intybus L.

ophthalmia, pulmonosis, tuberculosis, and typhoid. Flowers were once used in a collyrium to alleviate conjunctivitis (Grieve, 1974). CAUTION: Grieve (1980) mentions that, used in excess, it may result in loss of retinal vision or cause sluggish digestion. Chicory, as an additive to coffee, has caused dermatitis in workers handling the mix.

CIRSIUM SPP.   (ASTERACEAE) --- Thistles

DESCRIPTION: Prickly biennial or perennial herbs, often with a stout taproot and basal rosette. LEAVES: of the rosette somewhat dandelionlike, to as much as 18 inches long, often gray and variegated, with the teeth ending in sharp prickles; stem leaves one to the node, alternating along the stem, gradually smaller up the stem, spiny, without a clear leafstalk, the leaf base merging into and often embracing the stem. FLOWERS: in tight clusters, each cluster resembling a purplish burr, with spiny leaflike appendages surrounding the burr, the burr solitary, paired, or in small groups near the ends of the major stems. FRUITS: narrow, miniature, sunflowerlike nutlets with a dandelionlike parachute, the parachute not stalked like that of the dandelion and salsify.

DISTRIBUTION: Serious weeds, especially in pastures where cattle rarely graze them, avoiding the spines; also occasional in savannas, roadsides, or even swamps, requiring complete or partial sun, flowering from early summer until frost. *Cirsium arvense,* the Canada thistle, can reduce wheat yields 15–60%. More common in pastures is *Cirsium vulgare:* WA-ME-CA-FL. Zones 3–8.

UTILITY: To me, the most pleasant culinary appeal of the thistles I have indulged in is the chewy taproot on a hungry autumn afternoon. Those dug in pastures should not be eaten raw, because they may harbor disease. Other foragers have mentioned, with good reason, the edible pith of young stems, cooked young leaves, the stalks, pared of their spines and leaves, and the leaflike appendages around the burrs, cooked and consumed like globe artichokes. Some Canadian Indians have learned, independently as did I, that the taproots make good food, raw or cooked. But the taproot, like that of the evening primrose, is only good at the end of the first summer or at the beginning of the second summer of the biennial species. In the second summer, the biennials bolt, moving all that food upstairs to nurture the flowering process. The lilac-colored flowers sticking out of the spiny green structures below, when chewed, gum up to make a pleasant chewing gum substitute, as do flowers of sow thistle and dandelion. Thistle seed are a favorite with several seed-eating birds, like goldfinches. Canada thistle flowers are very attractive to honeybees. Delaware Indians used the bull thistle in their steam baths for rheumatism. Iroquois used it for cancer and hemorrhagic hemorrhoids. Navajo used it to induce vomiting. Ojibwa used it for stomach cramps. Canada thistle has been used to coagulate milk. Mohegan Indians used it in a mouthwash for infants and in a tuberculosis remedy for adults. Montagnais also use it for tuberculosis. Ojibwa used it as a bowel tonic. Various thistle species show up in Hartwell's *PLANTS USED AGAINST CANCER,* for cancer

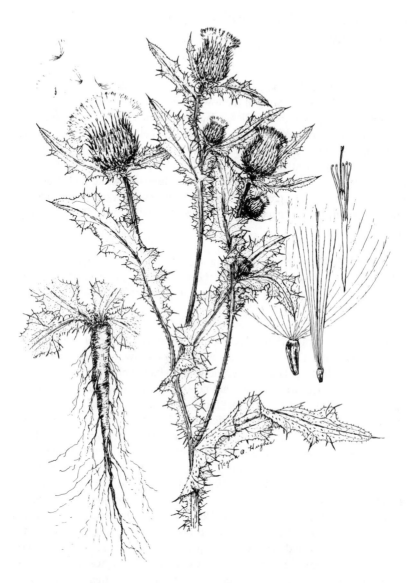

Cirsium vulgare Savi (Tenore)

of the breast and nose, edematous tumors, and scirrhus. WSSA lists over a dozen thistles, most of which can probably provide foodstuffs for the patient forager. CAUTION: The spines can even injure tough hands, and certainly tender mouth parts, inflicting such frightening words as ''papular urticaria'' and ''conjunctival nodules'' (Mitchell and Rook, 1979).

COMMELINA COMMUNIS L.   (COMMELINACEAE) ---
Asiatic Dayflower, Wandering Jew

DESCRIPTION: Smooth, hairless, juicy, clambering annual, rarely more than a foot tall, often rooting at the lower nodes, the roots fibrous. LEAVES: stalkless, broad, grasslike blades, emerging from tubular, flangelike structures, alternating or spiraling along the stem, the blades narrowly egg-shaped in outline, without teeth, but sometimes with small hairs at the edges of the leaf blade, tapered at the tip, with ca. 9–11 veins paralleling the margin. FLOWERS: few at the tips of the stem, blue and white (all blue in *C. virginica*), wrapped in a greenish, boatlike enclosure, open only 1 day. FRUITS: minute, green, turnip-shaped containers with usually 2 reddish to brownish seeds.

DISTRIBUTION: Often forming pure stands in moist garden plots, streamsides, and open woodlands, usually near homesites. Flowering as early as May in the Carolinas, June in Maryland, until frost. NE-NH-AR-AL. Zones 5–9.

UTILITY: Several writers allude to the edibility of several species as potherbs. I have eaten flowering shoots of this species raw. I also enjoyed tenderer parts of the plant cooked as a potherb. Not bad, not good. In India, both seeds and herbage are used as famine food and fodder. Japanese use the ephemeral flowers, we read, in making a special paper called Awobana paper. Javans steam young plants of *C. diffusa* as a vegetable, while Africans, Asians, and Hawaiians gather masses of the weed as fodder for their cattle (Morton, 1981). Latin Americans use various species for conjunctivitis, dermatitis, dysmenorrhea, enteritis, gonorrhea, kidney ailments, leukorrhea, malaria, and other venereal diseases (Morton, 1981). In China, the dayflower is used for abscesses, ascites, bleeding, boils, bug bites, colds, conjunctivitis, diarrhea, dysentery, dysuria, enteritis, fever, flu, gonorrhea, hypertension, laryngitis, malaria, mumps, snakebite, and tonsillitis. Duke and Wain (1981) list yet other maladies treated with various species of *Commelina:* buboes, burns, cachexia, cardosis, cough, cramps, cystitis, itch, leprosy, menorrhagia, nosebleed, ophthalmia, otitis, piles, tuberculosis, and uterosis. In spite of these long lists of supposed medicinal applications, *Commelina* is not a major medicinal species. Two *Commelina* species are mentioned in Hartwell's *PLANTS USED AGAINST CANCER*. Moerman (1986) mentions use of a Navajo species of *Commelina* in infusion as an aphrodisiac given to livestock. WSSA lists *Commelina benghalensis* L. (Tropical Spiderwort) and *C. diffusa* Burm. f. (Spreading Dayflower) as weedy species. I would not hesitate to sample either of these.

Commelina communis L.

CRATAEGUS SPP.   (ROSACEAE) --- Haw, Hawthorn, Thorn

DESCRIPTION: Deciduous shrubs or small trees, often with thorns at some of the nodes. LEAVES: one at the node, alternating along the stem, often varying very much on a given plant, even branch, broadly egg-shaped, lobed and/or toothed at the margin, 0.5–3 inches long, 0.3–2 inches wide, broadest at, below, or above the middle, pointed at the tip, but notched, squared, or pointed at the base, the leafstalk short to nearly as long as the blade. FLOW-ERS: one to few or many, often in malodorous clusters emerging from the angle formed by the leafstalk and the stems; sepals 5, united below; petals white, occasionally pink; stamens in 1–5 rows of 5 each; ovary 1- to 5-celled, immersed in the tube formed by the union of the sepals. FRUIT: a rounded, 1- to 5-seeded berry, some shade of green, orange, red, or yellow, often hanging on the tree late into fall, suggestive of a cross between a rose hip and an apple, but more the size of the rose hip.

DISTRIBUTION: A huge genus of many confusing species, covering the range of this small book, in intermediate forest and scrubs, neither real humid nor real arid, flowering mostly in spring, fruiting mostly in fall, and hanging on, but rarely 'til Christmas. Not knowing the taxonomist who can identify all these, I'm taking the easy way out and suggesting that the genus is mostly eastern, but that some species can be found in every state: WA-ME-CA-FL. Zones 3–9.

UTILITY: Jogging around the Beltsville Capitol of the USDA in fall, I find and consume what looks like ten or more species of *Crataegus* in a 1-hour lunch period. Leaf size, fruit size, shape, and color vary tremendously, as do the taste and grittiness of the hawthorn ''core''. It is easy to find ten of these eyeball species in the fall in an hour lunch period, and to enjoy eating half of them raw, the other half being either too gritty or too astringent. But it would take days or years, perhaps a lifetime, to identify them positively to species. Hawthorns and blackberries are better eaten than identified. Both can be eaten out of hand, or made into beverages or jams or jellies. Hawthorns are so important as medicinals as to have been featured as an illustration on the cover of issue number 22 of the trade journal *HERBALGRAM*. In that issue, herbalists Chris Hobbs and Steven Foster note that in medieval England, children ate the nutritious fruit. Fruits were also used in Russia to make wine. Chinese make jams, jellies, and wines from their species. At least a dozen American species are documented as food sources. ''Currently, hawthorn leaves and berries have no official regulatory status in the U.S.'' (Hobbs and Foster, 1990), but *C. oxyacantha* was once listed by the FDA as an ''herb of undefined safety''. I suppose undefined safety is better than no safety at

Crataegus crus-galli L.

all. CAUTION: Certain hawthorn species are famous for having a stimulant effect on the heart. I'm not sure I could tell the cardiotonic species from the native wild species. Hairs from *Crataegus* fruits can cause nodose conjunctivitis and the thorns are "particularly toxic to the eye". Corneal scratches led to loss of vision in more than half of 132 reported incidents in Ireland (Mitchell and Rook, 1979). WSSA lists 6 weed species.

CRYPTOTAENIA CANADENSIS (L.) DC. (APIACEAE)
--- Honewort, Wild Chervil

DESCRIPTION: Slender, hairless, branching perennials to 3 or perhaps 4 feet tall, from a cluster of fibrous roots. LEAVES: at the base compound, with 3 leaflets each, the leaflets narrowly egg-shaped, 2–6 inches long, 1–3 inches broad, mostly broader below the middle, pointed at the tip, with irregular saw teeth on the margins, basally tapered, with several laterals off the midvein; upper leaves smaller, but built on the same pattern, alternating on the stem, the leafstalk tending to form a sheath around the stem (as is typical of the family APIACEAE). FLOWERS: inconspicuous, in flat or rounded clusters at the tips of the main branches; sepals 5, united; petals 5, white or greenish white; stamens 5; ovary 2-celled, with one ovule in each cell, the ovary terminating in 2 processes. FRUIT: a narrowly ellipsoid, ribbed nutlet, round in cross section, 0.2–0.4 inch long, splitting lengthwise into halves.

DISTRIBUTION: Here and there common in meadows, pastures, and rich woodlands, mostly of the piedmont and lower mountains, flowering May to June, fruiting June to August, sometimes dying back before frost. MN-ME-OR-GA. Zones 3–8.

UTILITY: Roots are cooked like parsnips, and taste about as good (or bad, depending on your point of view). Young leaves, suggestive of chervil, or whole shoot tops are cooked as spicy potherbs or more often added to soups and stews. French use the herb in spring, in green soups, like chervil. Facciola (1990) states that young leaves, flowers, and stems are boiled and eaten as a potherb, chopped and added to salads, or used in green soups. He suggests scrubbing the roots and boiling for 20 minutes, serving with butter and parsley, or with a cream sauce. Stems are candied in sugar like angelica. One Oregonian was cultivating several acres of this herb for export to Japan, but he may have grown the Asian species. Japanese mound up their own species, very closely related, in winter, so that the shoots will be blanched in spring. Facciola says of the Japanese *C. japonica* that leaves and blanched leafstalks are eaten raw in salads and sandwiches, boiled, fried, added to soups, or used in egg dishes, tempura, or garnish. They often garnish it with sesame or soy sauce. Roots are blanched and sauteed in sesame oil, or they are boiled in concert with diced parsnips. Seeds are said to be useful, like aniseed, as a spice, or sprinkled on cookies, etc. Chinese use the decoction for colds, diarrhea, dysmenorrhea, glandular ailments, and rheumatism (Duke and Ayensu, 1985). A questionable entry in Duke and Wain (1981) hints that the aromatic herb might serve as a female tonic or aphrodisiac. CAUTION: I feel strongly that amateur foragers should stay away from mushrooms and this

Cryptotaenia canadensis (L.) DC.

family, the APIACEAE, or more colloquially, the carrot or umbel family. There have been too many fatalities due to mistaken identities, not always by the amateurs; sometimes the experts make fatal mistakes. Two foragers died out west in 1986, mistaking a poisonous umbel for a parsnip. The WSSA has not gotten around to listing this weed, but it's quite weedy. I could harvest a few pounds from shadier areas at Herbal Vineyard.

CYPERUS SPP. (CYPERACEAE) --- Chufa, Nutgrass, Nutsedge

DESCRIPTION: Greenish-yellow, weedy perennials, 1–3 feet tall when in flower, with triangular stems arising from horizontal rootstocks to which small tubers are often attached. LEAVES: long, narrow, grasslike (very narrowly sword-shaped), with several equal parallel veins traversing the length of the "sword", pointed at the tip, toothless on the edges, basally flared out to clasp the stem; most leaves basal, occasionally with a few stem leaves, one at the node, alternating on the stem, with a whorl of leaves flaring out beneath the flower cluster. FLOWERS: greenish to brownish, in complex branching clusters at the tip of the stem above the whorl of uppermost leaves, the minute flowers flattened, enclosed in greenish or brown scales; sepals and petals reduced or absent; stamens 2–3; ovary 1-celled, free of the scales, with 2 or 3 terminal processes. FRUIT: a small nutlet usually embraced by the scales, brown, lens-shaped or triangular in cross section.

DISTRIBUTION: Mostly weeds, especially common in dry to wet sandy fields, some species growing on the margins of streams and swamps, beginning to flower in late spring and early summer, often flowering and fruiting until frost. *Cyperus esculentus:* WA-ME-CA-FL. Zones 4–10.

UTILITY: The most famous species, *Cyperus esculentus,* called "chufa", has long been cultivated as a food and beverage plant, the tubers eaten as is, or ground up to make the beverage also known as "chufa" or "horchata", especially around the Mediterranean. As with many cultivated species, "chufa" has wild types of the same species or related species which give us less noteworthy food items, and are hence considered more weed than feed. de Vries (*Econ. Bot.* 45(1):27.1991) speaks of it as a weedy cultivar or a cultivated weed. The cultivated "chufa", originally from a Mediterranean climate, cannot long survive in the northeast, though thriving farther south. The wild form ranges to Canada, bearing smaller tubers, farther from the mother plant, decreasing their value as food, and increasing their liabilities as weeds. Once found, the tubers can be eaten, raw or cooked, or ground to make a beverage after soaking, or scorched and ground to make an "instant coffee". Elias and Dykeman (1982) suggest roasting the tubers until dark brown, grinding and brewing a tablespoon per cup. Toasted tubers are called "earth almonds" (BCW). Wills (WT1:2.1987, with cover pictures of nutsedge) gives a nice summary of the uses of nutsedges, mentioning their nutrients, allelochemicals, and medicinal compounds. Duke and Wain (1981) list abscesses, bladder ailments, boils, cancer, colds, colic, hemorrhage, stomachache, and ulcers as just a few of the ailments treated with just these two species, regarded as alterative, analgesic, antihistaminic, astringent, bactericide, carminative,

Cyperus rotundus L.

demulcent, diaphoretic, diuretic, emmenagogue, emollient, fungistatic, hypotensive, stimulant, and stomachic. Purple nutsedge can produce more than 2 tons/acre tubers, which is a lot of a tuber touted as aphrodisiac. There could be a lot going for two of the "World's Worst Weeds", *C. esculentus,* the Yellow Nutsedge, and *C. rotundus,* the Purple Nutsedge. WSSA lists 14 other weeds in the genus. It is easier to eat them than to identify them.

DAUCUS CAROTA L. (APIACEAE) --- Queen Anne's Lace; Wild Carrot

DESCRIPTION: Biennial herbs, forming a basal rosette and carrot-smelling taproot the first year, growing as high as 6 feet and flowering the second. LEAVES: of the rosette fernlike, very finely and complexly divided, the ultimate segments very narrow, pointed, the stem leaves smaller, but also complexly divided, with the leafstalk flared out and clasping the stem. FLOWERS: topping off the plant with a flat bird's-nestlike cluster of dozens or even hundreds of small white flowers, the clusters long stalked, the stalks arising between the upper leafstalks and stem, or on the opposite sides of the stem; sepals 5; petals 5; stamens 5; ovary 2-celled, embraced by the sepals, with 2 terminal processes. FRUIT: a flattened nutlet, ca. 0.2 inch long, with minute spines along some of the ribs of the fruit.

DISTRIBUTION: Common weed of old fields, waste places, perennial gardens, flowering as early as May and as late as frost. WA-ME-CA-FL. Zones 4–8.

UTILITY: If you don't believe that the Queen Anne's lace is the same species as the carrot, leave one of your carrot rows unharvested next year and let it go to flower. Like magic, the carrot has become the Queen Anne's lace. Like other biennials, the carrot draws most of the energy out of its roots if allowed to go to flower. And the roots of the flowering plant are withered, woody, and hardly worth the forager's effort to dig them. I have eaten them, but don't recommend them. They may be scorched and ground up as a poor person's instant coffee. Cattle, horses, and sheep graze the foliage. Facciola suggests that seeds are used to flavor soups and sauces, and that flower clusters are "French fried" as a gourmet dish. CAUTION: Some people suggest that the roots are poisonous. Regardless, amateurs should avoid them because of possible confusion with rather similar poisonous species like poison hemlock. Beginners beware! Poison hemlock can be confused with wild carrot. How well I remember an episode at Yogaville Virginia, when I started my weekend class by showing the students how similar were the edible Wild Carrot (*Daucus carota*), the medicinal Sweet Annie (*Artemisia annua*), and the deadly poisonous poison hemlock (*Conium maculatum*). I even showed them the purple spots on the poison hemlock stem. Still, two of my introductory students ate the poison hemlock within less than 4 hours after they had been warned about it. Fortunately, they had remembered my admonition to sample new things sparingly and had ingested only a bit of leaf. The plant they found in the field, 8 feet tall, didn't remind them of the small leaf I had used for demo that morning. Nervously, but trying to hide our nervousness, the nurse and I consulted the *CRC HANDBOOK OF MEDICINAL PLANTS* to see what

Daucus carota L.

antidotes were available, should they become symptomatic. Fortunately, they had ingested little more than homeopathic doses. Several informants from Pennsylvania have told me that the seed might dangerously serve as morning-after contraceptives. Such folklore has found experimental backing in some Asian Indian studies of mice, where extracts of the seeds prevented implantation of the egg following fertilization.

DIGITARIA SANGUINALIS L. (POACEAE) --- Crabgrass, Dewgrass, Large Crabgrass, Manna Grits, Twitch Grass

DESCRIPTION: Clumped, mat-forming annual herb, rooting at the nodes of the horizontal stems, 6–30 inches tall. LEAVES: one at the node, alternating along the stem, the blades long and grasslike, to 8 inches long, with a few parallel veins extending the length of the blade, the tip pointed, the edges toothless, but sometimes hairy, the base dilated to form a hairy, sheathing leafstalk that extends down the stem, enclosing it. FLOWERS: minute, stalkless, flattened along the long flat terminal branches (3–9) of the plant, each with about a dozen reduced flowers; sepals and petals not recognizable as such; stamens usually 3; ovary solitary, 1-celled, with 2 terminal processes, sometimes embraced by greenish to brownish ribbed "envelopes". FRUITS: small, flattened along the stems within their "envelopes", less than 0.1 inch long.

DISTRIBUTION: Common annual weeds of sandy soils, perhaps most frequent in poorly tended lawns and gardens and in old fields, roadsides, and waste places; flowering through summer up to the killing frost, which kills the annual mother, but ripens the dozens of seeds, thus assuring next year's weeds. Can be a serious weed in beets, corn, flax, peanuts, potatoes, sorghum, and soybeans, becoming resistant to the triazine herbicides. WA-ME-CA-FL. Zones 4–8.

UTILITY: Seeds, minute though they are, may be stripped off and used to make porridge, beverages (fermented to make beer), or toasted and ground to make flour. When cutting the fruiting plants in August with a hand scythe, I find the seeds congregate in large masses on the blade. These I have scraped off and cooked to make a poor man's oatmeal. I have no doubt that a good brewer could make a good brew from fermented crabgrass seed. A single plant may set as many as 150,000 seeds. Even in the temperate zone, it can produce 2–3 seed crops per season, seeding from early summer to frost. Seeds are substituted for rice in Poland and for grits elsewhere. Tropical species of *Digitaria* may yield more than 17 tons biomass per acre in a year. The plant has a folk reputation, for cataracts, debility, indurations, and scleroses. Grasses are easy to recognize as the family POACEAE, but are difficult to identify to genus, much less species. Almost all grasses are safe sources of cereals. CAUTION: Grasses and cereals may be moldy, especially in damp periods of the year. Black molds are particularly dangerous. Wilted grasses should also be avoided, as they may contain cyanide. Fruits of many wild grasses, as well as cultivated cereals, may have stiff hairs which can lodge in the throat, causing serious irritation. Such hairs should be removed somehow,

Digitaria sanguinalis L.

by winnowing or burning. Children sometimes exhibit grassitch following contact with crabgrass. According to Mitchell and Rook (1979), 10% of patients exhibited positive patch tests to crabgrass preparations. WSSA lists ca. 10 weedy species of *Digitaria*.

DIOSPYROS VIRGINIANA L.   (EBENACEAE) ---
Common Persimmon

DESCRIPTION: Deciduous trees, 45–60 feet tall at maturity, the bark dark brown and checkered. LEAVES: one at the node, alternating along the slightly zigzag stem, egg-shaped, usually broadest below the middle, pointed at the tip, without teeth or hairs, or with small hairs below, basally tapered to the short leafstalk which is usually less than one fourth as long as the leaf blade; midrib with several lateral veins arising like plumes of a feather. FLOWERS: 1 or few in small clusters in the angle of the leaf with the stem, often with some strictly female flowers (even strictly female trees); sepals 4(–5), fused below; petals 4(–5), cream or yellowish, fused below to form a cup; stamens ca. 16; ovary often 4- or 8-celled, independent of the sepals, with 4 terminal processes. FRUIT: an orangish-pink astringent berry, broader than long, 0.4–1.5 inches in diameter, with 3–8 brown, lens-shaped seeds.

DISTRIBUTION: Common, sometimes forming thickets, in fencerows, deciduous forests, and on the margins of coniferous forests; flowering May to June, but fruits not maturing or fit to eat until after frost. KS-CT-TX-FL. Zones 4–8.

UTILITY: Persimmons are one of the most abundant, easily obtainable and recognizable sugary foods available to us after frost. (The season of 1991 was highly atypical, with drought-induced ethylene reducing astringency before frost.) Then, and only then, they are good to eat out of hand, and are made into beers, breads, cakes, jams, molasses, pancakes, pies, preserves, puddings, vinegars, and wines. Some Indian tribes ground the dried fruits to make a breadstuff. Scorched seeds make a substitute (poor, in my opinion) for coffee. One old recipe for persimmon vinegar calls for 3 bushels of fruits with 3 gallons of whiskey (cheaper back in moonshine days) and 27 gallons of water. Even *SCIENCE* magazine (November 6, 1942) reports a vitamin C-rich leaf tea said to taste like sassafras. Reading that, I raced out and availed myself of some leaves, to make a beautiful chartreuse-colored tea, not sassafras-flavored. So I grabbed 10 leaves of sassafras, 10 leaves of spice bush, and 10 leaves of persimmon to make my Trinity Tea. Not bad, especially with lemon and sugar. The tree's flowers, though inconspicuous, are important bee forage. The dark brown, heavy wood, close kin to some tropical ebonies, has been used for golf clubs, shoe lasts, shuttles, turnery, weaver's shuttles, and other items requiring hard, smooth-wearing wood. Folk medicinally, persimmon has been used for cancer, diarrhea, dysentery, hemorrhage, kidney ailments, menstrual irregularities, sores, including sore throat, stomatitis, tumors, and uterosis (Duke and Wain, 1981). Cherokee chewed the bark for heartburn and used the plant in steam baths for biliousness and indigestion.

Diospyros virginiana L.

They took the bark tea for bilious problems, hepatosis, and toothache. The plant also entered their remedies for hemorrhoids and venereal diseases. From it, they made a syrup used for sore throat and thrush. Rappahannock specified bark from the north side of the tree for their thrush and sore throat remedies. (Of course there's likely to be more moss on the north side of the tree.) Recently, experimental mice proved oriental persimmon juice useful for hypertension and stroke (*Chem. Pharm. Bull.* 38:1049.1990).

ECHINOCHLOA CRUS-GALLI L. (POACEAE) ---
Barnyardgrass

DESCRIPTION: Coarse, annual herbaceous grass, usually less than 3 feet tall, occasionally up to 6, with no basal rosette. LEAVES: solitary at the nodes, alternating along the stem, grasslike, the blades 2–16 inches long, tapering to a point at the tip, with several parallel and nearly equal veins, toothless on the edges, basally tapered to a collarlike sheath nearly surrounding the stem beneath the point of attachment. FLOWERS: in complex terminal clusters, arising from the collar of the uppermost leaves, short-stalked, the clusters 2–12 inches long; sepals and petals barely recognizable as such; stamens usually 3; ovary 1, 1-celled, with 2 terminal processes, sometimes partially embraced by greenish ribbed "envelopes". Seeds small, oblong or ellipsoid, rounded or flattened in cross section, whitish brown when mature, less than 0.1-inch long.

DISTRIBUTION: Old pastures, feedlots, lawns, and other waste places, growing tall in organic situations, starting to flower in early summer with some flowering until frost, the fruits available to foragers from July to frost. A principal weed around the world in beans, beets, corn, cotton, potatoes, and rice, e.g. WA-ME-CA-AL. Zones 4–8.

UTILITY: Like other grasses, for example, corn, these smaller fruited grains can be eaten in the "milk" stage when green or in the "grain" stage after they have dried out a bit on the plant. Some Indian tribes parched the mature seed, grinding them up to make flours which they converted into breadstuffs. Japanese use ground seed in dumplings and macaroni. The seeds are reportedly used to adulterate fennel, according to The Wealth of India (CSIR. 1948–1976). Prolific, one barnyard plant can yield 1–2 million seed (WT4:918.1990). That may account for the development of cultivated millets (var. *frumentacea,* which was once called "billion-dollar grass" or Japanese millet). Facciola (1990) adds that young plants, stem tips, and hearts are eaten, raw or cooked. Barnyard grass may yield 4 tons/acre biomass, yielding as much as 1.5–3.5 tons of dry hay in 1.5–3 months in India, for example. Egyptians use the species in reclamation of saline soils. Conversely, as a weed, it can lower rice yields 1–2 tons/acre. Folklorically, *Echinochloa* is used for cancer, carbuncles, hemorrhage, sores, splenosis, and wounds and as a tonic and preventive. All the species of *Echinochloa* have wildlife value (like almost all members of the grass family POACEAE), being eaten by waterfowl (ducks, scaups, shovelers and teals); marshbirds (gallinules, rails, and snipes); gamebirds (doves, pheasants, and quail); and songbirds (blackbirds, cowbirds,

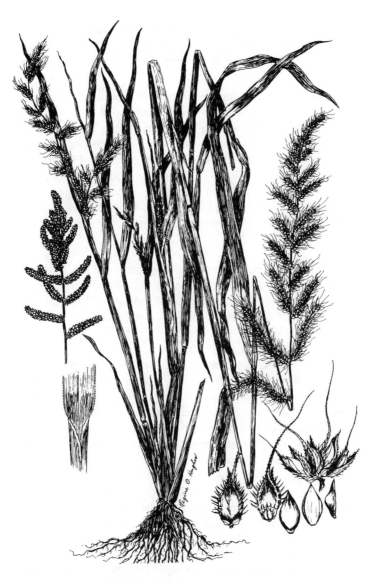

Echinochloa crus-galli L.

longspurs, and sparrows). Muskrats and rabbits eat the foliage. WSSA lists 7 weedy species of *Echinochloa,* all of which could provide food, fodder, and fuel (biomass). CAUTION: Occasionally, levels of nitrates might be high enough to cause problems in cattle grazers, not the humanoid grazer indulging in small tidbits.

ELAEAGNUS ANGUSTIFOLIA L.   (ELAEAGNACEAE)
--- Oleaster, Russian-Olive, Silverberry, Sugarberry

DESCRIPTION: Thicket-forming silvery shrubs or small trees, to as much as 30 feet tall, reduced twigs sometimes suggestive of thorns. LEAVES: willow- to elm-shaped, but without teeth on the edges, the blade with minute but distinctive silvery scales, with faint veins extending towards the edges from the midrib, 1–4 inches long, 1 inch or less wide. FLOWERS: small, cream-colored or silvery, solitary or in few-flowered clusters near the leaves, not much longer than the flower stalks, ca. 1/2 inch long or less, like a 4-lobed, stalked goblet. FRUITS: juicy, one-seeded, oblong, yellowish to pinkish or orangish fruits like miniature plums, but with miniature silvery scales.

DISTRIBUTION: Widely planted as an ornamental for reforestation on alkaline, minespoil, and sandy soils, the sugarberry tends to be a bit invasive. In Maryland, it is frequent in open alluvial floodplains. Flowering in summer, the fruits are usually available for the forager only a few weeks before frost. WA-MN-CA-MD. Zones 3–7.

UTILITY: The fruits, though astringent, are quite pleasant to eat. I remember gorging on these "sugarberries" as a grade school kid in Raleigh, North Carolina, having no idea they were alien introductions, thinking them a normal component of the old field flora down by the railroad track I haunted. Little did I know then that some of the olives of the Bible might be this silvery plant and not the silvery Mediterranean olive. Also known as Trebizond dates, the fruits are dried and powdered into an Arabian breadstuff. Fruits are also made into beverages which may be allowed to ferment. In Yarkland, a spirit is distilled from fermented fruits (CSIR, 1948–76). Facciola adds jelly, sherbert and wine to the products made from the fruits. Grosbeaks, pheasants, quail, and robins eat the fruits, and elk and presumably deer browse the twigs and foliage (Martin et al., 1951). Leaves of the plant are used as fodder for goats and sheep. Flowers are laden with nectar. Tolerant of alkaline soils and drought, it was introduced out west where the Soil Conservation Service regarded it as a good soil cover. The soft-to-moderately-hard wood is used for fenceposts and fuel (CSIR, 1948–1976). Some people are reported to use the seed oil, like olive oil, for bronchitis, burns, catarrh, and constipation. Flowers are used for fever, neuralgia, and aching burns, allegedly bringing people back from their deathbeds. The astringent leaves, used as fodder for goats and sheep, are used for enteritis and fever. WSSA also lists *Elaeagnus umbellata* Thunb. as a weed.

Elaeagnus angustifolia L.

ERECHTITES HIERACIFOLIA (L.) Raf. (ASTERACEAE)
--- American Burnweed, Fireweed, Pilewort

DESCRIPTION: Tall, bad-smelling annual herb to as much as 9 feet tall, the stems not hollow (as contrasted to the milky wild lettuce), the roots fibrous. LEAVES: one at the node, alternating on the stem, the upper willowlike, narrow, pointed at the tip, toothed at the margin, basally tapered to a short, ill-defined leafstalk sometimes clasping the stem; the lowermost sometimes lobed like a dandelion, 2–8 inches long, 0.2–2.5 inches wide; midribs with several lateral veins arising like the plumes of a feather. FLOWERS: minute, cream-colored to pinkish, aggregated into heads which the amateur might regard as individual flowers at the ends of the central axes and their main branches. FRUITS: small nutlets aggregated in the floral heads, about 0.1 inch long, brown, with a white, dandelionlike "parachute" at one end.

DISTRIBUTION: Common weed in old fields, roadsides, ill-kept lawns, waste places, and burned-over thickets; starting to flower in midsummer and usually persevering to frost. MN-ME-TX-FL. Zones 3–8.

UTILITY: I can scarcely undercut the understatement made by Peterson: "In Asia, the young leaves are eaten either raw or cooked. The strong flavor suggests that this is an acquired taste." My favorite foraging manual isn't much more complimentary: "In Asia the young tops and tender foliage are eaten, either raw or cooked. There is no reason, except the odor, to prevent our using it. Cooking may make it palatable to us" (Fernald and Kinsey, 1958). Facciola notes that Indonesians steam the flowering tops of *E. valerianifolia,* the Brazilian Fireweed, and serve them with rice. The tops are also eaten raw. Regarded folklorically as alterative, anodyne, antispasmodic, astringent, emetic, laxative, tonic, and vermifuge, the herb just doesn't sound too good. Still, in preparing for a trip to Amazonian Peru, I am looking at this plant, not just to fill a page in this book, but because it represents one of several annual weeds that occur in tropical America just as it does in the streets of Inner City Washington, D.C., where people often do not eat enough green vegetables. I predict I will see this species in Peru, and tell my classmates there how Venezuelans bathe in a boiled solution of the plant to dispel fevers. Salvadorans use the bitter decoction for cough. Morton (1981) mentions that it was once used in the U.S. for bloody diarrhea, eczema, and hemorrhages. WSSA also lists *Erechtites minima* (Poir.) DC. (Australian Burnweed) as a weed. CAUTION: Morton reports a few alkaloids in the plant that may be carcinogenic. I recommend that people avoid plants containing senecionine and seneciphylline.

Erechtites hieracifolia (L.) Raf.

ERODIUM CICUTARIUM (L.) L'Her.   (GERANIACEAE)
--- Alfilaria, Redstem Filaree, Storksbill

DESCRIPTION: Usually a winter annual, overwintering as an edible rosette of hairy compound leaves, clambering or almost erect from a taproot. LEAVES: of the rosette fernlike or yarrowlike (twice compound), 1–5 inches long, the leaflets arising like plumes off a feather's axis, the stem leaves one at the node, alternating on the stem, somewhat smaller. FLOWERS: in rather flat-topped clusters arising from the angles of the upper leaves with the stem, the stalk of the clusters longer than individual flower stalks; sepals 5; petals 5, pink or purplish; fertile stamens 5; ovary 5-celled, free of the sepals, with 5 terminal processes (styles). FRUIT: a long green pod 0.5–1.5 inches, the 5 sections spirally coiled when mature, opening from the base to the tip, the seeds ellipsoid, smooth, about 0.1–0.2 inch long.

DISTRIBUTION: Annual weeds, seemingly most common in sandy areas and lawns, at least in Maryland, also in waste places and cultivated fields, where it can be a troublesome weed; flowering mostly from March to June, with an occasional flower right up to frost or during warm spells in winter. CA-MA-TX-GA. Zones 4–7.

UTILITY: Tender leaves of the plant can be added to salads as is, or cooked. With butter, they make an acceptable potherb, especially if boiled in herbed and salted water and topped with lemon juice. Facciola (1990) notes that the leaves are added to omelets, salads, sandwiches, sauces, and soups. Further, he suggested that alfilaria leaves may be substituted in recipes calling for leaves of amaranth, beet, plantain, or sow thistle. Though the herb is a weedy introduction in America, the Blackfeet, Cahuilla, and Shoshone Indians quickly learned to gather it, eating it raw or cooked (Clarke, 1977). Costanoan Indians drank the leaf tea for typhoid. Navajo Indians used it for infections or following bites from members of the wildcat family. Europeans regard the plant as astringent, diuretic, emmenagogue, hemostat, oxytocic, and sudorific, using it folklorically for dropsy, dysentery, gonorrhea, hemorrhage, especially menstrual flux, rheumatism, sore throat, stomatitis, and uterosis. Amateurs may confuse this with some of the weedy members of the genus *Geranium,* most species of which are too astringent to be very palatable. Astringency in herbs and teas is often due to tannins, which can, under some conditions prevent cancer; under others, they can cause cancer. The wild geraniums are very good at stopping bleeding. Once on a live TV program at Herbal Vineyard, I rubbed abrasive comfrey leaves on an indolent ulcer (a nonhealing sore) on my shin. It started bleeding. Then I grabbed a wild geranium and rubbed it on the bleeding sore. It stopped bleeding immediately, right there on camera. By Monday, the sore had healed over, thanks to the comfrey a/o the geranium

Erodium cicutarium (L.) L'Her.

a/o the abrasion a/o the bleeding a/o the aeration and sunshine. CAUTION: Though the alfilaria is ready for consumption before flowering, amateurs should wait for the flowers, to avoid confusion with poisonous members of the carrot family. Looking over the European folkloric reputation as emmenagogue and oxytocic, I discourage pregnant women from indulging in this herb.

FRAGARIA VIRGINIANA Duchesne   (ROSACEAE) ---
Wild Strawberry

DESCRIPTION: Low, creeping, tufted perennial herbs with horizontal stems, often rooting at the nodes. LEAVES: with 3 leaflets (trifoliolate), alternating on the stem, but apparently arising singly or in tufts where the stems have struck root, the leaflets egg-shaped to willow-shaped, 1–5 inches long, 0.5–3 inches wide, pointed at the tip, toothed on the edges, rounded at the base above the leafstalk, which may be as much as 12 inches long; midribs of the leaflets with strong lateral nerves. FLOWERS: in long-stalked clusters, rising as much as 12 inches above the soil; sepals 5; petals 5, white; stamens about 20; ovaries numerous, spiraling around a conical axis, independent of the sepals. FRUIT: a bright red berry, egg-shaped, 0.2–0.8 inch long, the yellowish seeds outside the fruit (which represents the fleshy cone at the center of the flower).

DISTRIBUTION: Old fields and open margins of coniferous and deciduous forests, often forming large colonies, flowering March to June, closely followed by the fruits, closely followed by the frugivores. WA-ME-CA-FL. Zones 4–8.

UTILITY: Available so briefly in spring, the wild strawberries are my favorite spring fruits, and have been for half a century. They may be smaller, but they are tastier than the cultivated strawberry. Like cultivated strawberries, they can be served in dozens of attractive fashions. Indians dried or preserved the near-sacred fruits for use out of season. Their leaves, like those of blackberry and raspberry, make interesting astringent teas, pleasant to me when sweetened, but used once a month by many females for other purposes. The tannins mentioned under the alfilaria may contribute to that astringency. Early in 1991, it was announced that strawberries were a good source of ellagic acid, a known cancer preventive, related to the ellagotannins in conventional tea, *Camellia sinensis,* which also has some cancer-preventing reputation. Strawberries figure almost as heavily in the Cherokee legend of creation as corn. They keep strawberry preserves in the home to ensure happiness. Some of their dishes combine them into a strawberry johnnycake, corn muffins with strawberries. Strawberry johnnycakes could keep me pretty happy. Cherokee used a strawberry (leaf?) tea for dysentery and to calm the nerves. They also took it for bladder, kidney, and visceral ailments. Strawberry leaf tea contains so much vitamin C that Johns Hopkins University at first accused the late Euell Gibbons of faking his test results (spiking his samples?) (Squier, 1991). Moerman (1986) reports that Cherokee used the fruits to remove tartar from the teeth, a use showing up in some of the ''Food Farmacies''. Following along that same vein are reports that strawberries might be usefully eaten in

Fragaria virginiana Duchesne

cases of stones (Erichsen-Brown, 1979). Chippewa took the root infusion for infantile cholera. Micmac used it for an irregular menses. Ojibwa used it for stomachache. The yellow-flowered "snake strawberry", *Duchesnia indica,* equally weedy, has been described as "poisonous", but I think insipid would be a more appropriate adjective, having consumed quite a few "snake straw-berries" myself, with no great displeasure or pleasure.

GALIUM APARINE L. (RUBIACEAE) --- Bedstraw, Catchweed Bedstraw, Cleavers

DESCRIPTION: Clambering, angular-stemmed annual herbs, rough to the touch, the stems as much as 3 feet long or longer. LEAVES: whorled (6–8 around the node), willow-shaped, 1–3.5 inches long, 0.2–0.5 inch wide, broadest below the middle, pointed at the tip, toothless on the edges, but sometimes with marginal hairs, tapering to the stem, almost devoid of leaf-stalk, only the midrib prominent (other *Galium* species may have 3–5 veins extending from the base towards the tip). FLOWERS: in stalked, flat-topped clusters in the angles between smaller upper leaves and the stems; sepals 4; petals 4, white; stamens 4; ovary 2-celled and 2-lobed, immersed in the sepals, which are no longer apparent as the twinned fruit ripens. FRUITS: twinned, 1-seeded, globose, green to brownish or blackish nutlets, about 0.1 inch long, a bit broader than long.

DISTRIBUTION: Common weed, often forming clambering colonies, in waste places, old fields, pastures, usually in partially shady situations, but also in full sun; flowering April to June, the fruits closely following the flowers, sometimes persisting to frost. Near buildings in Maryland, one can find one or another species of *Galium* all winter long. WA-ME-CA-FL. Zones 3–8.

UTILITY: Younger shoots are said to constitute good potherbs, boiled about a quarter hour and topped with butter (I prefer margarine for health and gustatory reasons). Shoots can be boiled, cooled, and served with salads. As Facciola (1990) puts it, young leaves and stems are steamed and eaten as a vegetable or used in soups, stews, and Lenten pottage. I have eaten dozens of the gritty green fruits at a setting, and not experienced the emetic, laxative, or poisonous effects attributed to the weed. Belonging to the coffee family, cleavers are said to make one of the better coffee substitutes, used, for example, in Ireland (Uphof, 1968). Conversely, dried fruits are used as tea substitutes. Ripe fruits are slowly parched and ground to make a poor man's "instant coffee". (But really, foragers are spiritually rich, if financially poor.) Writing this in late May, I parched half a cup of seed in the toaster at 350°F for 10 minutes, then ground them up with mortar and pestle, and used the "instant coffee". Not bad! Remember that even real coffee is an acquired taste. I can't enjoy the real or the poor man's coffee without sugar. The dried plant is sometimes used as a tea. Jones (1991) describes the plant as "a delectable food, a superior drink, and a worthwhile herbal medicine". Chippewa used *Galium aparine* as a laxative and for dermatitis. Cowlitz women placed it in their bathwater to make them "successful in love" (Moerman, 1986). Fox Indians took the plant as an emetic. Iroquois used it for itch and

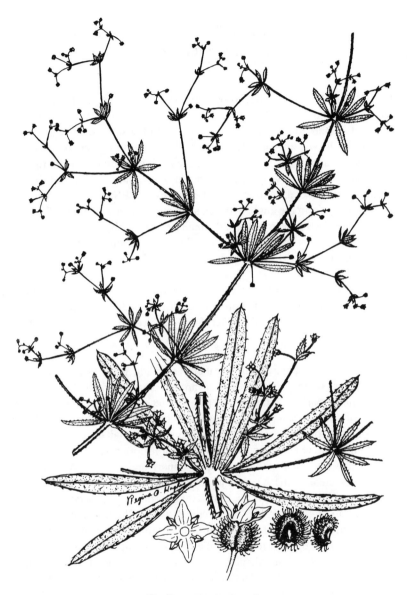

Galium aparine L.

poison ivy. Micmac used it for bloody sputum, gonorrhea, and kidney ailments. Ojibwa and Penobscot also used it as a diuretic for various urinary problems, kidney ailments, urinary blockages, and the like. Europeans regard the plant as alterative, aperient, apertif, depurative, diuretic, febrifuge, hemostat, refrigerant, sudorific, and tonic. CAUTION: Cowlitz Indians regard it as a poisonous plant. WSSA lists 4 *Galium* species as weeds.

GAULTHERIA PROCUMBENS L. (ERICACEAE) ---
Checkerberry, Teaberry, <u>Wintergreen</u>

DESCRIPTION: Aromatic, woody, evergreen, vinelike perennials, trailing along the ground, with short branches rising up to 8 inches. LEAVES: one at the node, alternating along the stem, but with the short stem telescoped to make all the leaves appear almost whorled, elliptic to egg-shaped, broadest below, at, or above the middle, 0.5–2 inches long, pointed at the tip, with minute teeth on the edges, basally tapered to the short stalk; midrib rather prominent, the laterals not so prominent. FLOWERS: solitary or in small clusters arising from the angle between stem and leaf; sepals 5, basally united; petals 5; stamens 10; ovary 5-celled, finally embraced by the sepals or nearly so, with a single terminal process. FRUIT: a small, pink or red, globose to ellipsoid berry, about 0.3–0.4 inch long and wide.

DISTRIBUTION: Acid thickets and deciduous and coniferous forests, sometimes abundant, but often quite rare, flowering June to August, fruiting in fall, the fruits persisting all winter if not gobbled up by foragers. ND-ME-AL-NC. Zones 3–7.

UTILITY: If you have smelled teaberry chewing gum, you know the delightful aroma of the wintergreen, an aroma shared with the cherry birch. Foliage and/or bark of either make pleasant teas. Some people suggest the addition of the aromatic leaves to salads, but they are too tough for my salads. This is one of the few edible fruits I can find on Christmas Day, along with the more common, but less tasty multiflora rose hips and partridge berries. Like the leaves, the fruits can impart their aroma to beers, teas, desserts (ice cream, in PA), and jellies or jams. For wintergreen pie (only where berries are abundant), macerate (saving some of the seeds for replanting), then mix 3 cups mashed winterberries with 1/2 cup sugar and 1/8 cup flour, starch, or meal, put in pie crust and bake 3/4 hour at 350°F. If you have been foolish enough to save some of last summer's milkweed chewing gum, you might even simulate the teaberry gum. Indians added the leaves to their smoking mixtures. Northern Indians revered the leaves almost as much as the Inca revered the leaves of coca, the source of cocaine. CAUTION: The medicinal methyl salicylate, common to wintergreen and cherry birch, can be poisonous if consumed in large quantities. Like other salicylates, methyl salicylate has figured as a treatment for rheumatism and arthritis, often as a counterirritant. Salicylates and other nonsteroidal anti-inflammatory drugs (NSAIDs) take 10,000–20,000 American lives a year, according to the trade journal *Chemical Marketing Reporter*. Often, these

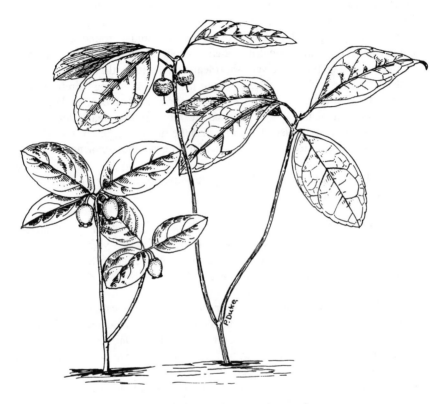

Gaultheria procumbens L.

fatalities were arthritics, treating their symptoms of painful inflammation, who developed bleeding ulcers as a result of their NSAIDs. Painfully aware of this as I take four ulcerogenic NSAIDs for the pain of a herniated disk, I am taking the roots of a Turkish weed, licorice (*Glycyrrhiza glabra*) for its reportedly protective effect against ulcers. (I have normal blood pressure. Hypertensives should be cautious about licorice, which reportedly raises blood pressure and lowers body potassium!)

GAYLUSSACIA BACCATA (Wang.) K. Koch
(ERICACEAE) --- Black Huckleberry

DESCRIPTION: Deciduous, thicket-forming shrub to about 4 feet tall. LEAVES: one at the node, alternating along the stem, 1–2.5 inches long, narrowly egg-shaped, broadest at or above the middle, rounded or shortly pointed at the tip, with the midrib sometimes sticking out at the point, toothless on the edges, tapered to the very short leafstalks, with dotlike glands on both surfaces; midrib with several less obvious lateral veins arising therefrom, like plumes from a feather. FLOWERS: green or reddish green, in branching clusters on last year's wood, i.e., below most of this year's leaves, the individual flower stalks usually shorter than the flowers; sepals 5, fused below into a bell-shaped tube; petals 5, also forming a tube; stamens 10; ovary 10-celled, united with the sepals. FRUIT: a black, waxy blue or white berry with ca. 10 seeds (to distinguish it from the superior blueberries, which have more than 10 seeds).

DISTRIBUTION: Dry woods and thickets, occasionally bogs, from the coastal plain to the mountains, flowering in May in Maryland, as early as March in North Carolina, the fruits maturing from June to September, but eagerly consumed by nonhuman foragers. ND-ME-LA-GA. Zones 3–8.

UTILITY: Berries are eaten, as is, or made into beverages, jams, preserves, or pies. In nonsurvival situations, one might prepare huckleberry muffins and pancakes. Huckleberries, like blueberries, bilberries, and cranberries of the same family (ERICACEAE), can be dried and made into pemmican, which can keep well into winter if properly prepared. Fruits are eaten by gamebirds (grouse, quail, turkey) and songbirds (catbird, crossbill, grosbeak, jay, oriole, tanager, towhee), bear, fox, and squirrels. Cherokee Indians chewed the leaves like chewing tobacco and Iroquois smoked the leaves. I don't know whether the leaf tea is supposed to be useful for high blood sugar and high blood pressure like those of blueberries and bilberries. Cherokee chewed the leaves for dysentery and tender gums, and took huckleberry tea for Bright's disease, colds, dysentery, and indigestion. Chippewa toke huckleberry tea as a blood tonic in spring and fall and for colds. Delaware also took it as a tonic, specifically for rheumatism. Iroquois took it to purify the blood, and for arthritis, colds, kidney ailments, rheumatism, tapeworm, and venereal disease. Lumbee took dwarf huckleberry tea for diabetes accompanied by slow-healing or nonhealing diabetic sores. Menominee took a tea of berries and leaves for rheumatism. Mohegans used huckleberry tea for kidney ailments. Ojibwa took the tea for rheumatism, and the Potawatomi for fever, lumbago,

Gaylussacia baccata (Wang.) K. Koch

and rheumatism. Shinnecock took the tea for kidney ailments (Duke, 1986; Moerman, 1986). In review, huckleberry tea has a widespread reputation for "incurable" rheumatism and arthritis. It does not have a reputation for causing bleeding ulcers like the NSAIDs. I'm glad I transplanted a huckleberry to my herb garden.

GLEDITSIA TRIACANTHOS L. (CAESALPINIACEAE)
--- Honey Locust

DESCRIPTION: Deciduous tree, rarely exceeding 100 feet tall, often with branched or unbranched thorns on the trunk or branches. LEAVES: one at the node, alternating along the stem, once- or twice-compound, the compound leaf with ca. 14–19 pairs of leaves arising along the midrib like plumes of a feather, or twice-compound, with the leaflets themselves subdivided; leaflets mostly 0.5–1.5 inches long, ellipsoid to narrowly egg-shaped, usually broadest below the middle, rounded at the tip, toothless on the margin, tapering to the short leaflet stalk; midrib with few laterals arising like plumes from a feather. FLOWERS: greenish, in clusters from the angle of this year's or last year's (as evidenced by leaf scars) leaves with the stem, sometimes borne on older branches or the main trunk; male clusters several in the angle; female clusters 1 per angle; sepals 3–5; petals 3–5; stamens 3–10; ovary 1-celled, free of the sepals. FRUIT: a large, flat, beanlike pod, leathery, green, turning brown, 3–16 inches long, 0.7–1.5 inches wide, the many brown, lens-shaped seeds embedded in a sugary brown pulp.

DISTRIBUTION: Common in deciduous forests and woodlots, along fencerows, once planted as cattle fodder, flowering April to June, fruiting July to November, the fruits often persisting after frost on the trunks of the trees. SD-CT-TX-FL. Zones 4–8.

UTILITY: The sweet, molasseslike pulp around the seeds can be used as a poor substitute for brown sugar and the scorched seeds can serve as a coffee substitute. Beer has been prepared by fermenting the sugary pulp, similar to the Biblical locust (better known to health food devotees as carob). One South Carolina beer recipe blends persimmon plus locust honey. But beer can be prepared from any salubrious item that has sufficient sugar and appropriate yeast species. It seems unlikely that the plant contains cocaine or atropine, as reported in Grieve's Herbal. I have submitted pods independently to two scientists who were unable to confirm the presence of cocaine. Tender green pods are cooked and eaten. Facciola adds that young seeds taste like raw peas. Trees are often used as ornamental or timber trees, providing wildlife with food. Cherokee used the pods for dysentery and measles and as an adjuvant to worm medicine, using the bark tea for indigestion and whooping cough. Creek boiled the branches, with thorns, for measles and smallpox. Delaware used the bark in cough remedies and as a tonic to purify the blood. Fox took it for colds, fevers, measles, and smallpox. Rappahannock used

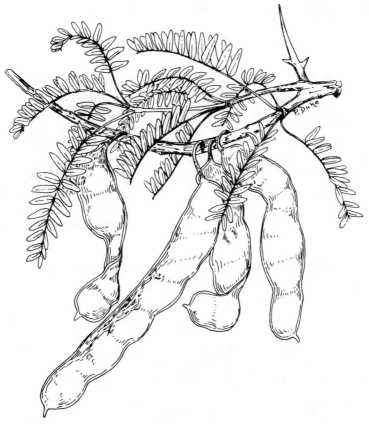

Gleditsia triacanthos L.

bark and roots in cold and cough remedies. The heavy play on smallpox evokes a limerick:

> The white man brought the smallpox disease
> That brought the red man to his knees.
> The red man, half joking, got the white man to smoking,
> And now the last laugh's a wheeze.

Not listed as a weed by WSSA. CAUTION: There is some overlap in the range of this species and the Kentucky coffee tree, whose pods may be mildly toxic if not processed. Statements in the literature about the wholesomeness of the coffee tree pods or those of the black locust should be viewed with wholesome skepticism.

HELIANTHUS TUBEROSUS L. (ASTERACEAE) ---
Jerusalem Artichoke

DESCRIPTION: Tall, herbaceous, often hairy perennials to as much as 12 feet tall, developing numerous strong tubers in fertile situations. LEAVES: 2 at the lower nodes (opposite), solitary and alternating higher up (alternate), varying from willow- to egg-shaped, usually broadest below the middle, drawn out into a point at the tip, with sawlike or irregular teeth on the edges, basally tapered to a short leafstalk, the blades 4–8 inches long, 2–4 inches wide, rough to the touch above (scabrous), often hairy beneath, with 3 veins arising at the base, the leafstalk 1/8–1/2 as long as the blade. FLOWERS: in tightly clustered, yellow, daisylike heads 2–5 inches broad, the central flowerlets minute. FRUIT: a small nutlet 0.2–0.3 inch long, with minute spinelets at the tip.

DISTRIBUTION: Woodland borders, open floodplains, meadows, and persistent in ill-tended gardens, flowering about late July and continuing until frost, at which time, apparently, much of the aerial "energy" is transported down to the roots. Thus the longer you wait (at least until frost), the bigger your harvest, in general. Once you know where the stands are, and the old flowering stalks may persist long after frost, you can find ample supplies all winter, at least when the ground is not frozen. It is my most copious producer of winter carbohydrate. WA-ME-CA-FL. Zones 3–9.

UTILITY: I find the juicy tubers to be suggestive of Chinese chestnuts raw, but suggestive of mush when cooked. Carefully peeled (the peel imparts an earthy flavor) and washed, sliced Jerusalem artichokes can be substituted for Chinese chestnuts in salads or other dishes that are not to be cooked. They do turn to an unattractive mush when cooked. The raw slices are nice for dipping into spreads as well. I can get a pound of tubers from one of my better plants, but I just can't eat a pound of tubers; they induce flatus. Were I poor, I think I could avoid overwinter starvation just with the untended artichokes around Herbal Vineyard. Diabetics might cultivate these roots because they contain inulin. Western Indians reportedly made a tea from the shoots for rheumatism (Duke, 1986). Moerman reported no medicinal uses for the artichoke, but compiled several for the closely related, seed-providing Common Sunflower, *Helianthus annuus*. Ironically, the sunflower, state flower of Kansas, has been declared a noxious weed in an adjoining state. Sunflower seeds are quite popular among health food "addicts" these days. Apache poulticed crushed sunflower plants to snakebite, a practice also observed by the Zuni. Dakota made a tea of the sunflowers for pulmonary affliction.

Helianthus tuberosus L.

Mandans used the seed oil cosmetically. Navajo used the sunflower for pre-
natal problems believed to have been caused by an eclipse of the sun. Paiute
used the sunflower for rheumatism, Pima for fevers and worms, Rees for
fatigue, Thompson for sores and swellings (Moerman, 1986). Seven species
are WSSA weeds.

HEMEROCALLIS FULVA L. (LILIACEAE) ---
Tawny Daylily

DESCRIPTION: Tall, hairless, perennial herb with a cluster of white to yellow rootstocks, often forming dense colonies. LEAVES: arising at the ground level, grasslike, 2–3 feet long, usually less than 1 inch broad, pointed at the tips, smooth on the margins, tapered and folded inward at the base, overlapping the folded bases of the leaves on the other side of the 2-ranked rosette, the veins running parallel from the base to the tip, only the midvein conspicuous. FLOWERS: few to several in long-stalked clusters reaching beyond the leaves, with a few leaflike blades below the stalks of the flower clusters; sepals 3; petals 3, orange; stamens 6; ovary 3-celled, with a single style and many ovules (undeveloped seeds). FRUIT: a 3-celled capsule, with several globose, black seeds (not developing in our Maryland escapees).

DISTRIBUTION: Common around old homesites, in meadows, old fields, stream banks, even in heavily shaded alluvial floodplain deciduous forests, flowering from May to July, apparently not maturing fertile fruit or seed. Flower buds, the safest food, are easily available from May to July; the more dangerous roots can be located year-round by the astute forager. After the Jerusalem artichoke, this is my most easily available tuber, all winter long. MN-ME-TX-FL. Zones 4–9.

UTILITY: Most books advocate the flower buds, in salads or soups. Orientals pick and dry them for year-round use. I enjoy 10–20 a day, during my "jogging foraging lunch period" raw, right off the stalk. They are also good boiled, pickled, or sauteed. Other books heartily advocate the young shoots and tubers as food, raw or cooked. Members of various foraging classes of mine have eaten them with impunity. The tubers give me diarrhea, raw or cooked. "Wildman" Steve Brill (n.d.) recommends cooking the bulbs (swollen yellowish roots) like potatoes, cutting a hole in the skin and then squirting out the contents, like toothpaste out of a tube. With no cautions whatever, Facciola adds that bulbs are eaten raw, baked, boiled, and creamed, or mashed and fried in fritters. Chinese consider the flowers anodyne, antianemic, antiemetic, antispasmodic, depurative, febrifuge, and sedative. They even give the flowers to deaden the pain of childbirth. Roots are used for dropsy, dysuria, jaundice, lithiasis, mastitis, piles, and tumors (Duke and Ayensu, 1985). CAUTION: People who suffer gout and may have taken colchicine may have become sensitized to that drug, and may suffer diarrhea, especially following ingestion of the roots. I believe this is due to colchicine or a similar alkaloid in the daylily roots. Indeed, were I suffering a gout crisis with no medicine available, I would immediately hazard the daylily bulb, expecting the same relief I expect from colchicine, following diarrhea. (But, planning to take the

Hemerocallis fulva L.

bulbs during my next attack of gout, I have gone over a year without one.)
Some people, not sensitized to colchicine, are also sensitive to the laxative
properties of the bulbs. Colchicine, formerly a drug of choice for gout, has
also been viewed as a cancer medication. Species of *Hemerocallis* have long
been used folklorically for cancer in China.

HORDEUM JUBATUM L. (POACEAE) ---
Foxtail Barley, Squirreltail

DESCRIPTION: Coarse, clumped or tufted perennial grasses to 2 feet tall, sometimes splayed out at the ground, with a dense, fibrous root system. LEAVES: solitary at the nodes, alternating along the stem, grasslike, the blades 8 inches long, tapering to a point at the tip, with or without minute hairs, rough to the touch above, with several parallel and nearly equal veins, toothless on the edges, basally tapered to a collarlike sheath nearly surrounding the stem beneath the point of attachment. FLOWERS: greenish to purplish, in tight, terminal, elongate, nodding clusters 2–5 inches long with long bristles (some to more than an inch long and often very annoying to cattle, lodging in their eyes, mouths, or nostrils), the terminal "flower" stalks arising from the collar of the uppermost leaves, sepals and petals barely recognizable as such; stamens usually 3; ovary 1, 1-celled, with 2 terminal processes, sometimes partially embraced by greenish ribbed "envelopes". FRUITS: yellowish, hairy "seeds", small, oblong when mature, about 0.1 inch long.

DISTRIBUTION: Old pastures, feedlots, roadsides, saltmarshes, and other waste places, flowering June to September or until frost, the fruits available to foragers from July to frost. WA-ME-TX-VA. Zones 3–6.

UTILITY: Like other grasses, for example, the cultivated barley, these smaller grains can be eaten both in the "milk" stage when green or in the "grain" stage after they have dried out a bit on the plant. Some Indian tribes (Nevada, Oregon, Utah) parched the mature seed, grinding them up to make flours which they converted into breadstuffs. Or parched barley seed can be used as a coffee substitute. There is some question as to which came first in the Middle East, primitive barley beers or breads. Some *Hordeum* species may yield 10 tons/acre biomass. Most barley species have wildlife value (like almost all members of the grass family POACEAE). Martin et al. (1951) list ducks and geese, gophers, ground squirrels, kangaroo rats, pocket mice, and prairie dogs as consumers of the seeds and leaves. Folklorically, wild species may possibly be used like true barley, for bronchitis, cancer, catarrh, chilblains, cholecystosis, cholera, cough, debility, fever, inflammation, measles, phthisis, pulmonosis, puerperium, sores, tumors, and urogenital ailments, and as a preventive for fever and gray hair. Chippewa use roots of *H. jubatum* in treating sties and inflammations of the eyelid. Costanoans use roots of an unidentified species in their medicine (Moerman, 1986). Eleven species of *Hordeum* are WSSA weeds. CAUTION: Long "spines" attached to the grains of some species can injure the throats of grazing animals, including humans.

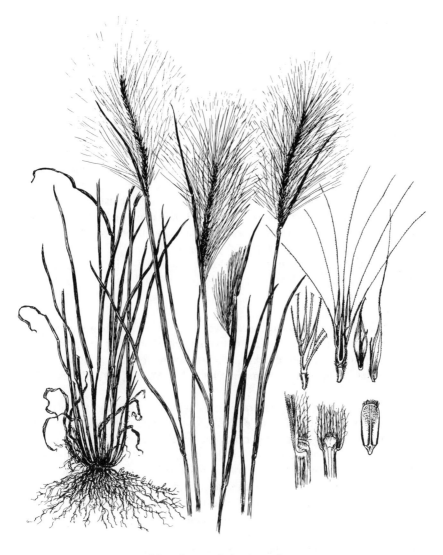

Hordeum jubatum L.

I have suffered this unpleasant experience on more than one occasion, when ingesting whole flower clusters of various grasses. Remove all spines from all foods before eating! Though not included in many poisonous-plant manuals, *H. jubatum* is believed among Navajo Indians to be poisonous (Moerman, 1986).

IMPATIENS SPP. (BALSAMINACEAE) --- Jewelweed, Touch-Me-Not

DESCRIPTION: Smooth, succulent, hairless, annual herbs to 10 feet tall, the stems hollow and translucent, the succulent roots reaching down like the "aerial roots" at the bottom of a corn plant, often reddish. LEAVES: one at the upper nodes and alternate, two and opposite at the lower nodes, the blades egg-shaped, 2–6 inches long, 1–4 inches wide, with a bluntly pointed tip, shallowly lobed on the edges, rounded at the base to the long leafstalk, often more than half as long as the blade, with several lateral veins arching from the midrib towards the edges. FLOWERS: 1–3 in small clusters in the angles between the upper leaves and stem; sepals 3, one expanded into a green sac; petals 5, partially united into a pealike arrangement, orange (*Impatiens capensis*) or cream to yellow (*I. pallida*); stamens 5; ovary 5-celled, free of the sepals. FRUIT: a turgid, green, "explosive" pod, 0.5–1 inch long, the seeds ellipsoid, 0.2–0.3 inch long, catapulted from the plant when the ripe "touch-me-not" is touched.

DISTRIBUTION: Marshes, meadows, swamps, alluvial bottoms, roadside ditches, ranging from complete sun to complete shade, flowering from May to June up to frost, then quickly knocked down, forming hollow, vaselike swellings at the uppermost remaining lobe. ND-ME-OK-GA. Zones 3–7.

UTILITY: Having read disparate views about the edibility of the plant, I elected to try it for myself. That's when I learned that most books were wrong about the leaf arrangement, describing the leaves as alternate (one at a node), which is only true of the top half of the plant, the half that botanists usually study in the herbarium. On the bottom half the leaves are opposite (two at the node). Being cautious, I took the jewelweed greens through two changes of water, freezing the water I strained off. Ice cubes made of jewelweed tea, like the foliage itself, are said to be good, either just after contact with poison ivy, or after the rash has already developed. Some even claim that it is preventive, like drinking the tea. (Some bold souls dangerously eat a little poison ivy, believing a little will bolster their immunity; science lends no credibility to such; three have told me of violent reactions following the ingestion of poison ivy; about as many told me of successes with such tomfoolery.) The greens, with a beautiful, rich-green color, tasted about as bad as spinach to me. I have eaten the pod with no displeasure, and have gathered the discharging seeds into a bag and found them edible in small quantities. They are appropriately described as tasting like butternuts. Seeds are eaten by mice, grouse, pheasants, and quail. Deer and rabbits graze the foliage and

Impatiens capensis Meerb.

hummingbirds visit the flowers. Still, I find no record of the Indian using this common plant for food. Should I view this lack of reported Indian food usage as a red flag? I discuss the jewelweed and enumerate several other presumed edibles that the Amerindians did not eat in *COLTSFOOT* 7(6). 1986. Clearly a weed in some pastured meadows, jewelweed is not listed by WSSA.

KOCHIA SCOPARIA (L.) Roth. (CHENOPODIACEAE) --- Belvedere, Burning Bush, <u>Kochia</u>, Mexican Fireweed, Summer Cypress

DESCRIPTION: Erect, often much-branched, annual herbs 2–7 feet tall, the stems rather succulent, hairy. LEAVES: simple, one at the node, alternating on the stem, 1–2 inches long, long and slender, pointed at the tip, with no obvious teeth or leafstalk, hairy. FLOWERS: small, greenish, short-stalked, in elongate simple or compound clusters toward the tip of the stem and major branches; sepals 5; petals 0; stamens 5; ovary simple, free of the sepals, with 2–3 processes at the tip. FRUIT: a small, papery, rounded pod, opening around the tip, each containing a single brown seed.

DISTRIBUTION: Common weed of the plains, often in pure stands, in cropland, old fields, pastures, waste places, usually in full sun, starting to flower in July, fruiting shortly thereafter, fruiting until frost. ID-ME-NV-MD. Zones 2–7.

UTILITY: Since I know this more as an ornamental than a weed in Maryland, I almost excluded this weed from this book. But, I was once involved with an entrepreneur who wanted to promote kochia as a marvelous food/fodder panacea for arid lands from Mexico to Saudi Arabia. Foragers and orientals have enjoyed the young shoots as a potherb. Orientals grind up the seeds as a cereal or foodstuff (Tanaka, 1976). The small seeds can be used in porridge or ground up like any other cereal to make a flour. Canadian studies suggest fodder yields of 3–5 tons/acre, making this a candidate for biomass production on alkaline, arid soils. New Mexico studies showed that the plant can produce 10 tons/acre biomass. Growing like a tumbleweed, the plant has been used for brooms. Recently it has been proposed to make artificial firewood or fuel briquets out of this and other tumbleweeds. *Kochia scoparia* is regarded as cardiotonic and diuretic. Navajo used one species of *Kochia* for sores. Chinese use their species for diarrhea, dysentery, dysuria, fever, gonorrhea, hernia, impotency, indigestion, incontinence, inflammation, intercostal neuralgia renitis, and urogenital disorders (Duke and Ayensu, 1985; Duke and Wain, 1981). Indians use *K. indica* for camel fodder, fuel, and as a cardiotonic (CSIR, 1948–1976). The weed has races that show tolerance to the herbicide triazine, especially along railroad rights-of-way where triazine has been used for a decade or more for total vegetation control. Furthermore, it produces allelochemicals which might discourage competing species. CAUTION: Although some people recommend it for forage

Kochia scoparia (L.) Roth.

production, others caution that in excess it can intoxicate animals, either due to alkaloids, nitrites, and/or oxalates (averaging 7% oxalates). The pollen is also reported to cause hay fever. Green-Molly, *Kochia americana* S. Wats., is also listed as a weed.

LACTUCA SCARIOLA L.   (ASTERACEAE) --- <u>Prickly</u> <u>Lettuce</u>

DESCRIPTION: Small to tall, spindly herbaceous summer or winter annuals or biennials to as much as 6 feet tall, exuding a white milk if broken, with a large taproot. LEAVES: often almost spiny, especially the crowded, larger, spiny lower ones, upwards single at the nodes, alternating up the stem, lower leaves dandelionlike, with prickly, toothed margins, pointed at the tips, to 12 inches long, 4 inches broad, leaves becoming smaller up the stem, the uppermost quite small and narrowly triangular; all grayish or bluish green. FLOWERS: yellow (sometimes reportedly drying blue), numerous, almost stalkless and scattered towards the tips of the upper branches, small, the stalks shorter than the flowers; sepals 5; petals 5; stamens 5; styles 2. FRUIT: a collection of minute, dark ''seeds'' bearing a silvery plume, parachutelike as in the dandelion and salsify, but smaller, the ''seed'' 0.1–0.2 inch long, flattened, flaring at the middle and pointed at the tip, brown to gray.

DISTRIBUTION: Barnyards, dumps, fencerows, old fields, pastures, roadsides, streamsides, vineyards, mostly in sunny, dry or moist situations. Starting to flower in June and flowering and fruiting until frost. WA-ME-CA-FL. Zones 3–8.

UTILITY: Like the ''tame'' lettuce, wild lettuces can have a bitter white milk which has been likened in appearance and sedative activity to opium. By the time they are flowering and hence identifiable, wild lettuces tend to be very bitter and tough. In the younger stages they can be used like lettuce, in salads, sandwiches, or soups, or steamed as a vegetable. Facciola notes that the seeds are the source of Egyptian lettuce-seed oil, a pleasant-flavored culinary oil. Among cultivated greens in this family we think first of lettuce, then of endives and witloof chicory, then even of the dandelions, now cultivated in New Jersey. When they are best in flavor, they are most difficult to identify, the wilder overwintering ones almost all having rosettes of leaves shaped like dandelions, with lots of lobing and teeth, and oozing a characteristic white juice (scientifically latex, often really containing rubber) when the leafstalk is broken. Flowers, when chewed, degenerate into a slightly bitter, cudlike chewing gum. As with several other ''milky'' plants, lettuces are folklorically believed to stimulate the flow of milk in humans. Bella Coola use *L. biennis* for diarrhea, heart ailments, hemorrhage, nausea, and pain. Following the doctrine of signatures, the Ojibwa used this milky plant for problems with human lactation. Cherokee took *L. canadensis* similarly, using it both as a stimulant and a sedative, and to allay pain. Iroquois took it for

Lactuca scariola L.

backache, puffy eyes, and kidney problems. Menominee rubbed the milk on poison ivy. Iroquois poulticed *L. pulchella* onto piles. Fox took the leaf infusion of *L. scariola* after childbirth to hasten the flow of milk. WSSA lists 7 weedy *Lactuca*, all relatively edible or inedible, depending on your point of view.

LEPIDIUM VIRGINICUM L.   (BRASSICACEAE) ---
Peppergrass, Poor Man's Pepper, Virginia Peppercress

DESCRIPTION: Small, herbaceous summer or winter annuals to as much as 30 inches tall. LEAVES: one at the node, alternating up the stem, sometimes crowded towards the base; upper leaves ellipsoid to willowlike, usually broadest above the middle, pointed at the tips, slightly toothed on the edges, tapered to a short leafstalk; lower leaves larger, pinnately lobed near the base; rosette leaves 1–4 inches long, less than 1 inch broad. FLOWERS: greenish or white, in elongate, rounded clusters at the tips of the main upper branch(es), small, the stalks about as long as the flowers; sepals 4; petals 4, sometimes absent in late-opening flowers; stamens 2, rarely 4; ovary 2-celled. FRUIT: a flat, round, 2-seeded pod, 0.1–0.2 inch long, slightly notched at the upper end, the seeds reddish brown, very slightly winged.

DISTRIBUTION: Weed of disturbed places, old fields, gardens, ill-kept lawns, flowering mostly April through June, but sporadically throughout the year as long as there is not frost. The herb is not easy to find in the heat of summer, however, availability peaking in spring, with a lesser peak in fall. WA-ME-CA-FL. Zones 4–8.

UTILITY: Like other members of the healthful and pungent mustard family (BRASSICACEAE), peppergrass can be added to salads, as one of the bitter herbs of the Bible, or cooked as a potherb. Some recommend adding the pods, without cooking, at the last minute, to soups and stews as a good seasoning. Facciola (1990) suggested chopping the pods into vinegar as a hot sauce. Lumbee Indians of the Carolinas were said to take peppergrass tea to restore the sex drive in the elderly (Croom, 1982). I enjoy nibbling on the tops in spring, stripping off leaves, flowers, and fruits with my teeth. In such indulgence of the plant, I found nothing to support folk claims that the herb is aphrodisiac. Matter of fact, there's one old Icelandic manuscript that indicates it "lessens the desire for women" (Erichsen-Brown, 1979). Fruits may be added to vinegar, both being improved in the process, at least to my taste. One leading foraging book says "Leaves contain vitamins C and A, iron and protein." I believe that is true of all green leaves (if you count the precursor of vitamin A, beta-carotene, as vitamin A). The mustard family, in addition, contains isothiocyanatelike compounds that have been suggested to prevent cancer. Geese eat peppergrass plants, finches eat the seed, rats eat the pods and leaves, and deer browse the plant. Cherokee used *Lepidium* poultices for croup and to draw out blisters. Houma took it with whiskey for tuberculosis. Menominee used peppergrass tea or bruised plants to apply to poison ivy (Duke, 1986; Moerman, 1986). Regarded folklorically as aperitive,

Lepidium virginicum L.

carminative, cicatrizant, diuretic, expectorant, laxative, and vermifuge, the peppergrass has been used for asthma, dropsy, gravel, hernia, rheumatism, ringworm, scabies, scrofula, scurvy, sores, and urogenital problems (Duke and Wain, 1981; Erichsen-Brown, 1979). WSSA lists 9 weedy *Lepidium*.

LYCOPUS SPP.   (LAMIACEAE) --- Bugle, Water Horehound

DESCRIPTION: Low, square-stemmed, nonaromatic perennial herbs to as much as 4 feet tall, spreading by slender runners from the base of the stem. LEAVES: two at the node, on opposite sides of the stem, 1–5 inches long, narrowly elliptic or egg-shaped, broadest below, sometimes at or above the middle, pointed at the tip, coarsely toothed on the edges, tapered to the base, the leafstalk short or absent, the midrib prominent, with a few lateral veins arching off like plumes of a feather. FLOWERS: small, white, tightly clustered around the stems where the upper leaves attach; sepals 4 or 5, basally united; petals 4, united; stamens 2; ovary 2-celled, with a single terminal process. FRUIT: a pair of 2-lobed nutlets.

DISTRIBUTION: Shady waste places, moist meadows, bogs, and woods, swamp forests, in deep shade or nearly full sunlight, flowering in summer (around June in Carolina, July in Maryland) fruiting to frost, by which time these, like many perennials, have transferred their food reserves to the roots for next year's growth (or for the forager). *Lycopus uniflorus* (Slender Bugleweed): WA-ME-CA-NC. Zones 3–7.

UTILITY: The gray or whitish horizontal roots, slightly resembling a grub, are said to be edible raw or cooked, or even pickled in wine vinegar. Washed and peeled, crisp tubers may be chopped up in salads as well. Chippewa, calling the roots what translated to "crow potatoes", dried and boiled them as food. Facciola (1990) notes that the crisp white tubers of *L. uniflorus,* suggestive of Indian cucumber root (*Medeola virginica*), are eaten raw in salads, boiled, pickled, or added to soups and stews. "Boiled a short time in salted water, they are said to be an agreeable vegetable, much suggesting the *crosnes* of European markets" (Facciola, 1990). Some plants, perhaps some species of *Lycopus,* may not form tubers, especially before frost, but those that do form tubers are well worth the digging. I have found tubers whenever I have dug around Herbal Vineyard. The characteristic stem, with its tight whorl of fruits, hangs on, albeit in a dry state, well into the winter so that the forager can find some plants on New Year's Day, if he wishes. Fox Indians used *L. americanus* as an analgesic in stomach cramps. Iroquois used the plant of *L. asper* as a children's laxative. Cherokee chewed the roots in the belief that it would make them more eloquent of speech. They also used it for snakebite, of themselves or of their hunting dogs (Moerman, 1986). Elsewhere regarded as aperitif, astringent, narcotic, and sedative, the *L. virginicus* is used folklorically for cough, diarrhea, diabetes, pulmonosis, and tuberculosis (Duke and Wain, 1981). WSSA lists 3 other weeds, *L. americanus*

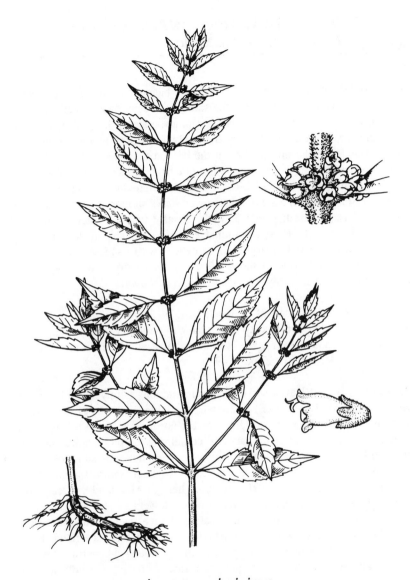

Lycopus virginicus

(American Bugleweed), *L. asper* (Rough Bugleweed), and *L. europaeus* (European Bugleweed), not listing the Virginia bugleweed, *L. virginicus,* which is the most common weedy species I encounter. CAUTION: Iroquois, though using the plants medicinally, report that *L. asper* and *L. virginicus* are poisonous. Uphof (1968) calls *L. virginicus* a mild narcotic.

MALVA NEGLECTA Wallroth (MALVACEAE) --- Common Mallow, Cheeses, Mallow

DESCRIPTION: Clambering, cold-tolerant, biennial or winter annual herbs, prostrate or ascending at the tips, the stems with star-shaped (stellate) hairs. LEAVES: one at the node, alternating along the stem, 0.7–2.5 inches long and wide, round or heart-shaped, shallowly lobed and toothed around the edges, rounded at the tip, notched at the base at the leafstalk, slightly hairy on both sides, with 5–9 veins radiating out from the leafstalk, the leafstalk longer than the leaves. FLOWERS: in few-flowered clusters from the angles of the leafstalks with the stems, the flower stalks shorter than the leafstalks; sepals 5, basally fused; petals 5, pink, basally fused; stamens numerous, united at the base to form a staminal column (rather characteristic of the family MALVACEAE); ovary 8- to 15-celled, with 8–15 styles. FRUIT: a round pod-like collection of the 8–15 one-seeded hairy "cells" separating from each other at maturity, the seeds blackish brown.

DISTRIBUTION: Weed of gardens, ill-kept lawns, pastures, and waste places, flowering as early as April, sporadically until frost. The foliage, and sometimes sporadic flowers, are available almost all winter in protected situations. WA-ME-CA-FL. Zones 3–9.

UTILITY: Boiled leaves serve as a mucilaginous (slimy like okra of the same family, MALVACEAE) vegetable, available 9 months of the year, or all year around nice warm buildings as far north as Maryland. The flower buds can be dropped in pickle jars still containing the pickle juice, but with no pickle. Flowers are pleasant enough to eat, and not quite so mucilaginous as the pods and leaves. Thus you make one of Father Nature's pickles. (The true caper and the clove are flower buds; here in Maryland you can make pickled flower buds of the daylily and other flowers as well.) Young fruits might be pickled as well or added to salads. Leaves can be added to soups, like okra, as both seasoner and thickener. Western Indians extended their pinole with mallow leaves during hard times. Young green fruits, the so-called "cheeses" or "biscuits", are good, raw or cooked. Facciola adds that ground fruits of *Malva parviflora,* the Little Mallow, with a few leaves for color, can be used for a creamed soup which resembles pea soup. He adds that the root decoction of *M. neglecta* may be substituted for egg white in meringue recipes. WSSA lists 8 weeds in *Malva,* most of them possibly edible. CAUTION: Mitch notes (WT4:693.1990) that horses, sheep, and cattle have reportedly suffered intoxication after indulging in fresh mallow. Muscular tremors called "shivers" or "staggers" have been induced by experimental feedings in Australia. Malvalic and sterculic acids in the plant can induce pink "Easter egg" yolks in chickens ingesting the mallows. Levels

Malva neglecta Wallroth

of nitrate that might be toxic to grazing cattle have been reported from *M. parviflora.* So far, the male contraceptive gossypol, widely distributed in the Malvaceae, has not been reported from *Malva,* but I am submitting material for analysis. Pythagoras did comment that ingesting mallow "reduced the passions" (WT4:693.1990).

MEDICAGO LUPULINA L.   (FABACEAE) --- <u>Black</u> <u>Medic</u>, Nonesuch

DESCRIPTION: Annual, biennial, or perennial herb to as much as 2 feet tall, the stems branching, taproot shallow. LEAVES: one at the node, alternating on the stems, "cloverlike" (composed of three similar leaflets), the leaflets egg-shaped, to 1/2 inch long, 1/2 inch broad, usually broadest above the middle, rounded or slightly notched at the tip, finely saw-toothed on the margin, tapered to the base, with numerous lateral veins arching out towards the margins, the leafstalk of the compound leaf often longer than the "leaf". FLOWERS: minute, yellow, pealike, in long-stalked, tight clusters close to the upper leaves; sepals 5, united basally; petals 5; stamens 10; ovary 1-celled. FRUIT: a small, 1-seeded hairy to hairless coiled pod. Seeds minute, bean-shaped, greenish to orangish brown.

DISTRIBUTION: Planted in pastures and volunteering in fields, pastures, and roadsides, usually in dry to moist sunny situations, flowering as early as April in the Carolinas, May in Maryland, black medic is of little interest to foragers except as the "poor man's alfalfa", setting seed on until frost. WA-ME-CA-FL. Zones 3–8.

UTILITY: Those of you who like alfalfa sprouts might want to try burclover or medic sprouts as well. Any of the *Medicago* species can be sprouted, as can any of those of *Trifolium* (clovers). You might get a few interesting compounds in the process. Legume isoflavones seem to be estrogenic and are believed by some NCI scientists to prevent cancer. Even the seeds are said to be edible, but that's a lot of work for a little seed. California Indians apparently ate the seed. They can be ground up as meal. Facciola (1990) notes that the leaves were used as a potherb. For many uses of the various *Medicago* species, readers are referred to the *HANDBOOK OF LEGUMES OF WORLD ECONOMIC IMPORTANCE* (Duke, 1981). Since most of our *Medicago* are imports, they have little Amerindian folklore. According to Moerman (1986), Costanoan Indians poulticed heated alfalfa leaves onto the ear for earache. Elsewhere, alfalfa is folklorically regarded as anodyne, bactericide, cardiotonic, depurative, emetic, emmenagogue, and lactagogue. Perhaps the estrogenic compounds increase the flow of milk. Alfalfa and other species of *Medicago* are used in folk medicine for such ailments as arthritis, boils, cancer, dysuria, fever, gravel, heart ailments, scurvy, and tumors (Duke and Wain, 1981). Alfalfa seeds are regarded as emmenagogue and lactigenic (Duke, 1981). CAUTION: *If* alfalfa sprouts and canavanine are linked to lupus, as some scholars claim, perhaps these sprouts might best be left off your menu, or ingested cautiously, in moderation if at all. Saponins in alfalfa might cause the breakdown of red blood cells, if grazed in excess. Nitrate

Medicago lupulina L.

concentrations greater than 0.2 per cent can be damaging to livestock and presumably grazing foragers who ingest too much alfalfa. Under certain conditions, the plant may generate cyanide (Duke, 1981). WSSA lists 4 other weeds in *Medicago*, the <u>Spotted Burclover</u>, *M. arabica;* the <u>Little Burclover</u>, *M. minima;* the <u>California Burclover</u>, *M. polymorpha;* and <u>Alfalfa</u> itself, *M. sativa.*

MORUS SPP.   (MORACEAE) --- Mulberry

DESCRIPTION: Deciduous trees to 60 feet tall or taller, the superficial roots exuding a little milk when broken, often brightly colored with yellow, orange, or purplish red. LEAVES: one at the node, alternating on the stem, sometimes lobed, sometimes mitten-shaped, sometimes unlobed (sharing this rare trait with sassafras and some crabapples), more often unlobed, smooth, and hairless in *Morus alba* (White Mulberry), more often lobed and rough and hairy on the veins below in *M. rubra* (Red Mulberry), blades broadly egg-shaped to round, pointed at the tip, toothed along the edges, notched or rounded at the base where it attaches to the long leafstalk, usually with 3–5 veins arising here and arching, often asymmetrically, towards the tip. FLOW-ERS: in separate male and female flowers or even different trees, in inconspicuous conelike clusters where some leafstalks meet the stem; sepals and stamens 4 in the male flowers, which hang 1–3 inches; sepals 4, the many ovaries 2-celled in the shorter female clusters. FRUITS: an aggregated fleshy cluster of fused female flowers, but resembling a white, pink, red, purple, or black dewberry, the colors highly variable, like the taste.

DISTRIBUTION: Common weed trees in disturbed coniferous and deciduous forests and old fields, especially along fences and under wires where frugivorous birds perch, flowering from March to May, the fruits ripening from June to July, and then gone for the year. *Morus rubra:* MN-NY-TX-FL. Zones 4–8.

UTILITY: Most of us probably ate mulberry fruits as children. There's not much to confuse them with, and they certainly are copious during their short fruiting periods, often making purple circles on the roads beneath the trees. Dried fruits can be ground up as flour. If you put your mulberries into a pot of water preparatory to cooking them into preserves or fruit juice, you may lose your taste for them. Thousands of minute insects called thrips scurry to the surface of the water. (Thrips may be a main source of vitamin B-12 for fruitarians or vegetarians.) Confusingly, the color of the fruits is not important in distinguishing the species, which seem to hybridize. Asian writers speak of the cooked young shoots of our commoner white mulberry as a good vegetable. Such shoots are cooked with rice, added to stews, or substituted for tea. One American author cautions that unripe fruits and raw shoots of the red mulberry "contain hallucinogens". I find the raw leaves more palatable than other tree leaves I have sampled; I have not found them to be hallucinogenic, however. Authors who seem sure they know the differences between the species have not convinced me. Paradoxically, Cherokee Indians used the *M. alba* bark both to loosen one up and tighten one up (as a laxative and to check dysentery; Moerman, 1986). Alabama Indians used *M. rubra*

Morus spp.

roots when passing yellow urine. Cherokee used *M. rubra* bark as an anti-dysenteric, laxative, purgative, and vermifuge. Creek use the roots as an emetic and for urinary problems and weakness. Delaware used the bark as an antibilious cathartic or emetic. Fox, like the Chinese, regarded mulberry as a panacea. Rappahannock rubbed the tree sap onto ringworm (Moerman, 1986).

NASTURTIUM OFFICINALE R. Brown
(BRASSICACEAE) --- Watercress

DESCRIPTION: Hairless, succulent, perennial aquatic herbs, sometimes 3 feet long, fixed in the water, rooting at the nodes. LEAVES: one at the node, alternating on the stem, often with white roots opposite the leaves, the leaf blades irregular, dandelionlike, 1–6 inches long, 0.5–2 inches broad, with a midrib running the length with 2–5 lobes on either side and flared out at the rounded or slightly pointed tip, the lobes not usually toothed. FLOWERS: in long, many-flowered clusters to 6 inches long at the stem tips; sepals 4; petals 4, white; stamens 6; ovary 2-celled. FRUIT: a long pod 0.6–0.8 inch long, with small seeds in 2 rows.

DISTRIBUTION: Occasionally forming large colonies in slow, but clear streams or brooks, especially in limestone areas, in full sun or partial shade, flowering April to July, and fruiting shortly thereafter on until frost. Available all year in Maryland. WA-ME-CA-FL. Zones 4–10.

UTILITY: An extremely nutritious herb, according to many sources, making a good salad, au natural, or garnish. Excellent soups can be made from watercress, using the same recipes you might use for sorrel or cabbage soups. Some modern restaurants use cress as a garnish, culinarily more attractive to me than the usual parsley. Since we cannot be sure our streams don't harbor harmful germs, we should only eat watercress after cooking or sterilizing. At my farm, I have a springfed, yearlong stream to which I have successfully transplanted watercress, just grabbing a mess from the third creek upstream from mine, and throwing it into mine. I haven't had my stream checked, but I would not eat or drink from it, without boiling the water or watercress. I even boiled a few of the crayfish one time. Crayfish and watercress make good potmates. Cooking the watercress as a potherb will certainly lower chances of getting disease organisms with your dose of watercress, and it will lower vitamin content, too, unfortunately. Recent USDA studies suggest that beta-carotene might survive the parboiling. One recipe calls for frying the cress in oil with ginger. Seeds can be germinated to produce delectable sprouts, or they can be ground up to make a "water mustard". Facciola (1990) suggests blending the herb with butter. Adding an authoritative oriental voice to the food farmacy fad, Leung (1984), echoed by Heinerman (1988), notes a carrot/watercress soup for canker sores or oral blisters. Cantonese make watercress soup with pork for dry cough and throat, or too much phlegm in the throat. Watercress has found its way into herbal cosmetics recommended for correcting skin blemishes and freckles. Hartwell's *PLANTS USED AGAINST CANCER* mentions folkloric use of watercress for boils, carcinomata, tumors, warts, and wens. Spanish-speaking New Mexicans, who call it "berros",

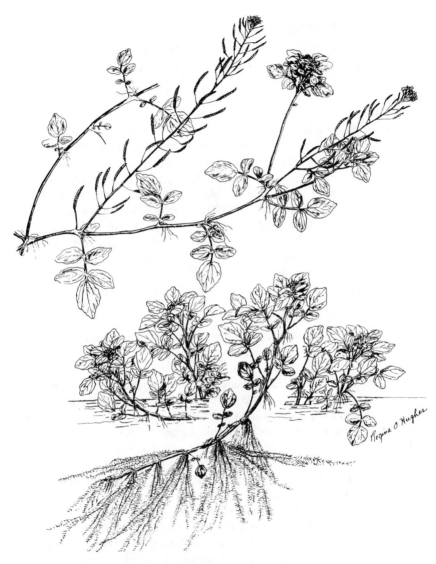

Nasturtium officinale R. Brown

use it for heart trouble (Duke, 1986). Costanoan Indians use cold watercress tea for fevers and for kidney and liver ailments (Moerman, 1986). Mahuna use it for cirrhosis, gallstones, and "torpid liver". For a triple threat against cirrhosis, blend it with dandelion flowers, milk thistle seed kernels, and tofu. Other ailments treated folklorically include asthma, baldness, bronchitis, eczema, flu, goiter (might be counterindicated), hepatitis, impotence (Jones, 1991), polyps, scabies, scurvy, and tuberculosis (Duke and Wain, 1981).

NELUMBO LUTEA (Willd.) Persoon
(NELUMBONACEAE) --- <u>American Lotus</u>, Duck Acorn,
Water Chinquapin, Water Nut, Water Lotus

DESCRIPTION: Obligate aquatic herbs, the leaves sticking out of the water unlike the floating leaves of the waterlily, with a horizontal rootstock sunk in the bottom mud. LEAVES: one at the nodes of the rootstock (actually an underground stem), alternating thereon, but all aiming for the air; blades round, depressed in the center, where the leafstalk is attached on the underside at the heart-shaped notch, umbrellalike, 4–28 inches wide, toothless, the leafstalk to 3 feet long or longer. FLOWERS: yellow, 5–12 inches in diameter, lifted out of the water on long flower stalks, 3 feet or more long; sepals and petals intergrading, 20 or more, the outermost green, grading into the yellow petals; stamens numerous; ovaries aggregated into something like an inverted showerhead with an ovule in each hole. FRUIT: consisting of the enlarged "showerhead", facing up, with the seeds about 0.3–0.5 inch in diameter.

DISTRIBUTION: Aquatic in ponds or sluggish streams, rarely in freshwater inputs to saline estuaries, flowering in early summer, fruiting until frost. MN-MA-TX-FL. Zones 4–8.

UTILITY: Some American Indians, e.g., the Omaha, like Asian Indians, tended to regard the water lotus as sacred. Indians, like the Chinese and well-trained foragers, also knew that once you know where this plant grows, you could always get Father Nature's handout. Tanaka (1976) notes that the entire plant is edible. Even in winter, the starchy rootstocks, up to 50 feet long, can produce copious food, once unearthed from the cold bottom of the pond. Swellings along the rootstock are said to be especially good, boiled or baked. Baked, they are said, perhaps euphemistically, to taste like sweet potatoes. The seeds, persisting for a while after frost, are almost uncrackable, but, like the Chinese water lotus, provide a starchy foodstuff, reducible to flour, once you have parched and cracked the seeds. Indians shelled the seeds, green or ripe, and added them to meat soups. Green seeds, edible raw or cooked, are said to taste like chestnuts. Indians used the leaves as a potherb. Any tender part of the plant apparently can be safely examined as a potential food source. If the flowers of our species are floated in soups as orientals do with *Nelumbo nucifera*, I don't know about it. Neither Duke (1986) nor Moerman (1986) reported Amerindian medicinal uses for the American species, but Duke and Ayensu (1985) and Duke and Wain (1981) report a litany of uses for the Asian water lotus, *N. nucifera,* whose seeds are said to remain viable for centuries. Considered antidotal (to mushroom poisoning), antiemetic, bacteristatic, demulcent, deobstruent, diuretic, hemostatic, stomachic, it is believed by the Chinese to prevent premature ejaculation and to improve virility.

Nelumbo lutea (Willd.) Persoon

The very hard seeds are eaten to preserve health, the cotyledons or seed leaves said to "purify the heart, permeate the kidneys, strengthen the virility, blacken the hair, make joyful the countenance" (Duke and Ayensu, 1985). Approaching my 62nd birthday, 3 days after April Fools', 1991, perhaps I should indulge these fountains of youth.

NUPHAR LUTEUM (L.) Sibth. and Sm. (NYMPHAEACEAE) --- Spatterdock, Yellow Pondlily, Yellow Waterlily

DESCRIPTION: Obligate, aquatic, perennial herbs, the leaves often lifted above the water, with cylindrical rootstocks horizontal in the mud. Elliot (1976), unable to find both ends of these huge rootstocks, speculates they may extend from one side of the pond to another. LEAVES: one at the nodes of the submerged rootstocks, alternating thereon, round to egg-shaped, 4–16 inches long, hairless above, hairy or hairless below, rounded at the tip, toothless on the edges, with a heart-shaped notch at the juncture with the leafstalk. FLOWERS: large, solitary, raised above the water; sepals 5–14, greenish, the inner tinged with color; petals numerous, yellow, grading into the numerous stamens; ovary 6- to 35-celled with 6–35 terminal processes. FRUIT: an egg-shaped berry 1–2 inches broad, with numerous yellowish to brownish, egg-shaped seeds about 0.1–0.2 inch broad.

DISTRIBUTION: In lakes, ponds, sluggish streams, estuaries, and swamps, flowering as early as April and fruiting until frost. ND-ME-TX-FL. Zones 4–8.

UTILITY: Indians used the thick, fleshy tubers, raw, roasted, made into breads, or boiled with meat. Modern writers emphasize, however, that the roots aren't sensational food items, even after several changes of water. Elliott (1976), describing the flavor of the roots as ''baked swamp'', wonders if other authors of other foraging books had eaten the roots or only passed on the bibliographic echoes. I agree that they taste like baked swamp. One writer even compares them to sheep's liver. Indians are said to have ground the seeds and added them to soups as thickeners. I can't vouch for claims that the seeds can be toasted and popped like popcorn, but that would certainly be an improvement over baked swamp. Elliott (1976) notes that squaws dove for the roots, or raided muskrat houses where the muskrats had stored the roots. The Indians would repay the muskrats with other foods, just as they repaid field mice whose stores of hog peanuts had been commandeered. Facciola says the leaf stalks are edible, and the flowers are made into teas. Unlike the American *Nelumbo,* the American *Nuphar* has evolved an Amer-indian folk repertoire. Menominee poultice the dried, powdered root on cuts and swellings; the Micmac to bruises and swellings; Ojibwa to cuts, sores, and swellings. Penobscot poulticed the leaves rather than the roots on swollen limbs. Potawatomi poulticed it onto inflammatory ailments. Rappahannock poulticed the leaves onto boils, fever, and inflammation. Sioux applied the

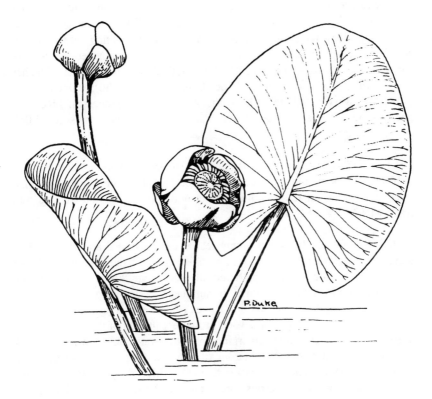

Nuphar luteum (L.) Sibth. & Sm.

powdered, dry rootstock as a styptic. Thompson Indians took cold spatterdock tea for internal aches and pains, poulticing the leaves onto cuts and sores. Hartwell's *PLANTS USED AGAINST CANCER* mentions spatterdock for cancer of the testes and uterus. The Western species had an equally colorful folk medicinal role, for asthma, backache, bone ailments, chest pains, lung hemorrhage, and rheumatism (Moerman, 1986). WSSA lists 3 weed varieties under *Nuphar*.

NYMPHAEA ODORATA Ait.   (NYMPHAEACEAE) ---
Fragrant Waterlily

DESCRIPTION: Obligate, aquatic, perennial herbs, from a forking root-stock paralleling the mud's surface at the bottom. LEAVES: one at the node, alternating along the rootstock, 3–12 inches broad, floating, green above, sometimes purplish below, round, untoothed, hairless, notched at the base for the long leafstalk. FLOWERS: long-stalked, but floating, 2–6 inches broad, white or pinkish; sepals 4; petals numerous (17–30); stamens many (35–100); ovary of 12–35 cells, with 12–35 terminal processes. FRUIT: a nearly spherical, many-seeded berry, maturing underwater, the seeds 0.05–0.1 inch long.

DISTRIBUTION: Pools, ponds, sluggish streams, sometimes a bit dwarfed in blackwater bogs, more frequent in the coastal plain than in the mountains, starting flowering in June, or July farther north, flowering and fruiting until frost. MN-ME-TX-FL. Zones 4–8.

UTILITY: Flower buds and young leaves can be boiled and eaten. Older leaves are used to wrap other foods for baking. As with the other waterlilylike plants, the rootstocks can be gathered all year, if you know where they are and what they look like in winter. In summer you can tell the waterlilylike plants apart roughly as follows: (1) leaves round, emerging from the water, with the stalk coming out of the back of the leaf instead of the bottom (water lotus, *Nelumbo*); (2) leaves emerging from the water but with the leafstalk attached at the notch in the bottom of the leaf (pondlily or spatterdock, *Nuphar*); (3) leaves and flowers floating (waterlily, *Nymphaea*). The seeds, once extracted from the berries, are ground and used as flour. Foster and Duke (1990) add that a mixture of lemon juice and waterlily root, was once used cosmetically to remove freckles and pimples. Waterfowl eat the seeds and roots, while cranes and gallinules eat the stems, roots and seeds. Beavers, moose, muskrat, and porcupines browse the plants (Martin et al., 1951). Chippewa used the dry powdered root to treat oral sores. Micmac used the roots to treat colds, glandular ailments, grippe, and swollen limbs; they also used the leaves for cold and grippe. Ojibwa used the roots for coughs and tuberculosis. Penobscot poulticed mashed leaves onto swollen limbs. Elliot (1976) notes that, as a douche, the root decoction is highly recommended against vaginal infections. He adds that the root decoction is used to treat diarrhea, as a wash for sore eyes, and as a gargle for sore throat. Hartwell's *PLANTS USED AGAINST CANCER* mentions folk use of this species for cancer, ulcerated cancers, hot, inflamed, or painful tumors, uterine cancer,

Nymphaea odorata Ait.

and whitlows. WSSA lists 7 weed species and/or varieties in *Nymphaea*. CAUTION: Foster and Duke (1990) caution that in large doses, the waterlily may be toxic. The widespread food and medicinal uses of these 3 water weed genera indicate that there may have been much confusion about them in the literature, even though they are rather distinctive.

OENOTHERA BIENNIS L.   (ONAGRACEAE) --- Common Eveningprimrose, Evening Primrose, King's Cureall

DESCRIPTION: Annuals or winter biennials (my term, I think, for those useful plants that germinate one year, make a big taproot for winter, flowering and dying the second year) to 8 feet tall in their second year, from the conspicuous rosette and taproot of their first year. LEAVES: of the rosette long, narrow, and willowlike, 3–9 inches long, to 1 inch wide, the midrib often reddish, with little or no leafstalk, pointed or rounded at the tip, toothless, wavy, or minutely toothed at the margins; stem leaves one at the node, alternating along the stem, progressively smaller up the stem, more pointed than the basal leaves. FLOWERS: nearly stalkless in elongate clusters at the top of the plant, the lowermost opening first, lasting only one night and part of the next day, 0.7–2 inches wide; sepals 4, united below; petals 4, separate, yellow; stamens 8; ovary 4-celled, united with the sepal bases, topped with a single stalk with 4 terminal processes. FRUIT: a long, dry, 4-sided pod, ultimately brown, 0.5–1.5 inches long, opening and curving back from the tip, with many minute seeds.

DISTRIBUTION: Lawns, roadbanks, old fields, and pastures, flowering as early as June, the first fruits ripening in August, the rosettes available to the forager from August on until late in autumn or winter, like the seeds. A good forager can find rosettes and last year's seeds even in winter and spring. OR-ME-AZ-FL. Zones 4–8.

UTILITY: Seeds, source of the magic gamma-linolenic acid (GLA), are available most of the colder months of the year, and can be sprinkled like poppyseed onto rolls, cornbread, toast, or salads, or just popped into the mouth for a quick GLA fix (only if the seeds are chewed and subsequently digested; whole, unchewed seeds probably mostly pass through undigested). The taproots, known as German rampions, are easily available in fall. I have gathered a pound in less than 15 minutes. I enjoy the taproots raw, with their turniplike afterbite, or cooked. Facciola compares the roots with parsnip or salsify, but I think they are more like rutabaga or turnips. Europeans boil the roots for 2 hours, then French-fry, or eat them as is, or buttered and salted and peppered. Roots are also suggested for pickling. There's no part of the plant I have not eaten: flowerbuds, flowers, cooked green pods, boiled leaves, etc., but I find the seeds and the taproots the most pleasant to the taste, as is, and processed. Petals are rather unremarkable, but not unpleasant as raw foods. Facciola suggests their use for salad garnishes or pickles. Moerman (1986) notes that the Cherokee used evening primrose tea for obesity, and poulticed warm "rampions" onto piles. Iroquois also used the roots in a pile

Oenothera biennis L.

treatment, and for boils and laziness. They rubbed chewed roots onto athlete's muscles in the belief that it improved strength. The Potawatomi were the only ones Moerman reported to use the seeds medicinally. GLA and evening primrose oil have made headlines lately as health food items. Unfortunately, the product(s) have been confiscated several times lately by the FDA, for making claims not substantiated to the FDA's satisfaction. Consuming the seeds for a decade, I have suffered no premenstrual syndrome yet. WSSA lists 5 weeds under *Oenothera*.

OXALIS SPP. (OXALIDACEAE) --- Sourgrass, Wood Sorrel

DESCRIPTION: Low, fibrous-rooted annual herbs or perennials, some-times bulbous. LEAVES: all from the root (in white- or pink- to purple-flowered species) or one at the node (in yellow-flowered species), alternating on the stem, cloverlike, with 3 notched leaflets arising at the tips of the long leafstalks, the leaflets so deeply notched at the tip as to appear heart-shaped, with one main vein going to each lobe of the heart, toothless on the edges, the leafstalk usually much longer than the leaflets. FLOWERS: solitary or in clusters of 2–9; sepals 5; petals 5, yellow, white, or pink; stamens 10, usually of 2 sizes; ovary 5-celled, free of the sepals. FRUIT: a green to tan 5-celled pod, with 2 to several reddish seeds in each cell.

DISTRIBUTION: Common weeds (the yellow-flowered species) in lawns, fields, pastures, etc., or herbs (the pink- or white-flowered species) of con-iferous or deciduous forests, flowering as early as March farther south, and flowering until frost. *Oxalis stricta* (Yellow Woodsorrel): WA-ME-CA-FL. Zones 4–9.

UTILITY: Most foraging books speak of the edibility and the pleasant sour acid taste of the leaves, reminding us of the dangers of consuming too many. I find the green fruits even more pleasant than the leaves, calling them wild pickles. Kindscher refers to them as ''little bananas''. I used to eat at least 10 a day for lunch on my jogging-foraging lunch period. Leaves and fruits are chewed to alleviate thirst. Wood sorrel is said to be good in salads, with watercress and wild onion, and has been suggested as a spice for beaver, muskrat, and porcupine meat. I have also enjoyed the flowers and small bulblets at the base of the purplish-flowered woodland species. Facciola sug-gests the flowers as a garnish for salads. Kiowa Indians chewed the leaves to alleviate thirst when on long trips (Uphof, 1968). Potawatomi Indians cooked the tubers with sugar to make a dessert. Rabbits and deer browse the foliage; buntings, doves, grouse, juncos, larks, quail, and sparrows eat the seeds. Upland gamebirds (doves, grouse, and quail) also eat the leaves and bulbs. Algonquin considered the wood shamrock, *O. acetosella* as an aph-rodisiac, while other groups use it for cancer (Duke, 1986). Cherokee used the Creeping Woodsorrel, *O. corniculata,* for blood disorders, cancer at its onset, chewing on it for mouth sores and sore throat. For hookworm, they used the tea orally and topically; for vomiting, they took a cold tea. Iroquois took the tea of the European Woodsorrel, *O. europea,* for cramps, fever, nausea, and just as a mouth freshener (Moerman, 1986). Seneca Indians boiled

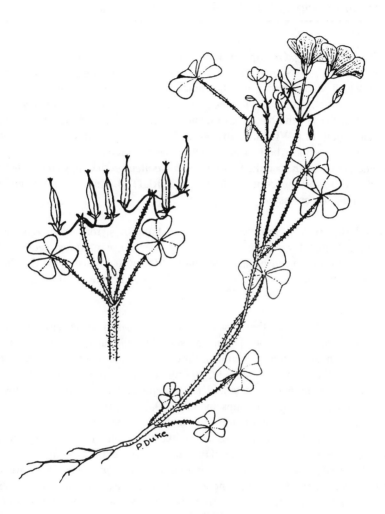

Oxalis sp.

Oxalis in bear grease to make a cancer salve (Erichsen-Brown). CAUTION: Like spinach, rhubarb, sorrel, etc., this plant contains oxalic acid, which, in large quantities can cause problems. Milk or a tea of wood ashes or anything high in calcium might serve as an antidote. WSSA lists 6 weedy species, all potential foraging foods.

PERILLA FRUTESCENS (L.) Britt.   (LAMIACEAE) ---
Beefsteak Plant, Perilla, <u>Perilla Mint</u>, Chiso

DESCRIPTION: Green, purple, or mottled annual mint, to as much as 3 feet tall. LEAVES: egg-shaped, long-stalked on opposite sides of the stem, with 4 to 7 veins coming off the middle vein and arching out towards the margins, which have many rounded teeth, tapered or rounded to the leafstalk, pointed at the tip, the blades 2–6 inches long, 1–4 inches wide, the stalk 1–3 inches long. FLOWERS: Numerous and small, very short-stalked, scattered along the uppermost branch or branches, with 5 greenish to purplish sepals around the purplish, pealike flowers, with 4 pollen-producing stamens tucked inside the flowers surrounding the nutlets. FRUIT: 4 minute, seedlike nutlets which function as seeds.

DISTRIBUTION: Barnyards, fields, old homesites, roadsides, volunteering in unkempt lawns, sometimes taking over horse pastures; seeming to thrive in full sun to dense shade. Starting to flower in August in the Carolinas and Maryland, fruiting up to frost. Resembling basil, the perilla is not quite so sensitive to frost. IA-MA-TX-FL. Zones 4–8.

UTILITY: Here's another case of one man's poison being another man's food. Around Washington, D.C., one sees this cultivated in window boxes or gardens near oriental restaurants. Orientals use the leaves, e.g., in sushi, and pickle the seeds in marinade. Young seedlings they eat with raw fish (sushimi). Salted leaves they add to tempura and tofu. Immature flower clusters are used to garnish soups and tofus. Older clusters are fried. At Herbal Vineyard one year, perilla completely took over a big stretch of the horse's pasture behind the barn. Horses don't like the plant, and may experience respiratory ailments if overexposed thereto. Japanese extract a compound from the plant that is another famous nonnutritive sweetener, many times sweeter than sucrose. I sometimes add the leaves to herb teas, but won't recommend that. It does not sweeten the tea. Recently I received an interesting book *WHY GEORGE SHOULD EAT HIS BROCCOLI* (Stitt, 1990), praising the health benefits of alpha-linolenic acid (ALA), on the heels of my *"Orthomegalomania"* (Duke, 1989) and Rudin's *THE OMEGA-3-PHENOMENON,* all praising the virtues of ALA. Perilla seed may be a better source of ALA than flax seed. While it is not exactly GRAS, I suggest that this weed could become one of tomorrow's health foods. Could it possibly be the ALA in chiso garnishes rather than the soy products that lay behind the epidemiological observations that orientals have less breast cancer than occidentals, *until* they embrace an occidental diet? Japanese studies show that one compound (xanthine-oxidase inhibitor) in the leaves has antigout activity equalling that of my allopurinol. Chinese use the antiseptic, antitussive seeds folklorically for

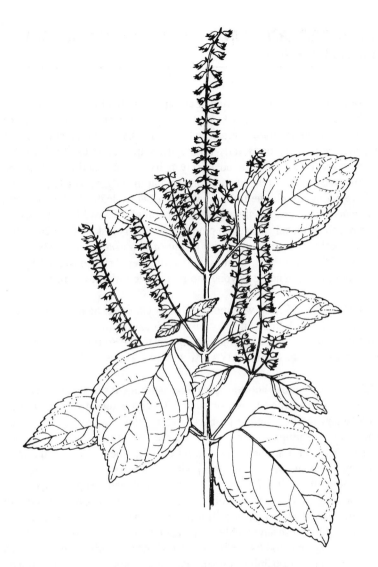

Perilla frutescens (L.) Britt.

asthma, cough, and premature ejaculation. Leaves share many properties with the seeds, used to ward off colds, and as an antidote to crab and fish poisoning (maybe that's why it is so important in sushi). With the plant regarded folklorically as alexiteric, anodyne, antiseptic, antitussive, diaphoretic, expectorant, fungicidal, pectoral, sedative, and stomachic, it also figures in folk remedies for bronchitis, cholera, cold, cough, flu, headache, malaria, nausea, rheumatism, and uterosis.

PHRAGMITES COMMUNIS (Cav.) Trin. (POACEAE) --- Common Reed, Reed

DESCRIPTION: Tall, erect, tufted or clumped bamboolike perennial grass, from a knotty horizontal rootstock; to 15, rarely 18 feet tall, and nearly an inch in diameter; stems grayish green, hairless. LEAVES: basal and then one at the node, alternately or spirally arranged on the stem, the blades 6–24 inches long and to 2 inches broad, rarely hairy at the junction of leaf and stem, sometimes rough to the touch (scabrous) at the edges, which are sometimes sharp enough to cut the forager; with no obvious leafstalk and no teeth; clasping where the leaf joins the stem; veins parallel. FLOWERS: numerous, tawny or purplish, closely spaced in plumose terminal clusters up to 24 inches long, 6 inches broad, sometimes with a ring of hairs on the stem above the last leaves. FRUITS: small, possibly not setting regularly, obscured by the surrounding hairs, extending beyond the minute, leaflike floral parts.

DISTRIBUTION: Weedy wetland grass of bogs, creek banks, moist roadsides, and swamps, often forming extensive stands, tending to be more common to the north, with *Arundo* more common to the south. Flowering and fruiting mostly July to October. WA-ME-CA-FL. Zones 3–7.

UTILITY: Only young shoots are palatable as forage for cattle and foragers, and it can be hard-earned starch or sugar. The reed reportedly sometimes exudes a sugary manna. Young rootstocks and stalks, before flowering, are sometimes sweet enough to serve as sugarcane substitutes. Such reeds may also be sun-dried, ground, and beaten into flour; the finer siftings of the flour are moistened to heat over the fire as a poor man's marshmallow. The sugary sap was used by the Paiute Indians as an expectorant and analgesic with lung ailments, a usage also reported from Africans. Rootstocks provide food all year round, roasted or boiled like slim potatoes. Apache used the roots for diarrhea and other stomach troubles. The decoction of the rootstock is also taken for earache and toothache. Emerging shoots in spring can be boiled like asparagus spears or bamboo sprouts. Partially unfolded leaves are cooked as a potherb or used as a tea substitute. The grains can be used as a poor man's cereal, searing off most of the fluff around the seeds. Amerindians did not even try to hull the seeds, but ground the whole grain into a nutritious, high-fiber meal. Reed has been used as a cellulose source for rayon manufacture. Wherever these aquatic plants grow, one finds them used for simple construction, for fencing, trellises, roofing, basketry, and mattings. It is such a good source of biomass (to 25 tons/acre) that it is often proposed as an energy source. Stems serve for fuel, fiber, musical instruments, and thatchery. Variegated and otherwise more handsome cultivars are often grown as ornamentals. Reportedly alexiteric, diaphoretic, diuretic, emetic, refrigerant,

Phragmites communis (Cav.) Trin.

sialogogue, stomachic, and sudorific, the plant is used folklorically for abscesses, arthritis, breast cancer, bronchitis, cancer, cholera, condylomata,
cough, diabetes, dropsy, dysuria, fever, flux, gout, hematuria, hemorrhage,
hiccoughs, jaundice, indurations, leukemia, nausea, pulmonosis, rheumatism,
sores, stomach, thirst, and thyroid. CAUTION: Until one is sure how to
distinguish this species from *Arundo,* one should not overindulge in the plant.

PHYSALIS HETEROPHYLLA Nees   (SOLANACEAE) --- Clammy Groundcherry

DESCRIPTION: Branching upright or clambering annuals or perennials, sometimes with star-shaped hairs, the perennials often with horizontal rootstocks. LEAVES: one at the node and alternating, or rarely two on nearly opposite sides of the stem, egg-shaped, usually broadest below the middle, pointed at the tip, shallowly or deeply toothed on the edges, or toothless, tapered to the leafstalk; midrib with several lateral veins arising like plumes on a feather, but not nearly so closely spaced. FLOWERS: greenish yellow, one each in the angles formed by the upper leaves with the stems; sepals 5, green, almost completely fused to form a tube, becoming a balloonlike covering when mature; petals 5, united (fused to their tips), yellow, often with a dark spot; stamens 5; ovary 2-celled. FRUIT: a round, many-seeded berry, at first green, later yellowish or brownish orange or red, enclosed in the greenish ''balloon'' which later becomes straw-colored or brown, at or after frost.

DISTRIBUTION: Here and there common in sunny situations, old fields, meadows, pastures, waste places, flowering in May or June, but the fruits not completely ripening until after July 4, like tomatoes in Maryland, but hanging on and edible right on up to New Year's Day, at least. WA-ME-ID-FL. Zones 3–9.

UTILITY: The ripe berries, rich in vitamins C and A, are edible as is, or stewed like tomatoes, or made into jams and/or jellies. Indians apparently ate most species of *Physalis,* sometimes grinding them with raw onions, chili peppers, and coriander seed. Dried fruits are ground into a breadstuff. Indian children used to pop the inflated balloonlike structures on their heads in play. Some Indian names for the plant translate to ''forehead pop''. Fruits are also eaten by grouse, pheasant, quail, and turkey, as well as mice, possums, skunks, and wildcats. Iroquois Indians washed with the leaf and root tea for burns, scalds, and venereal disease. They took the tea as an emetic when they suffered bad stomachaches. Omaha, Ponca, and Winnebago used *P. virginiana* root decoctions for headache, stomachache, and to dress wounds (Moerman, 1986). Could the Fox Indian usage of *P. virginiana* for dizziness foreshadow the application of solanaceous alkaloids for vertigo? CAUTION: Most parts of this plant, perhaps even the unripe fruits, are inedible, perhaps even poisonous. The leaves of some species have the unpleasant aroma of jimsonweed (*Datura*) containing poisonous, but medicinal alkaloids. Two of those alkaloids are very important in wars like the recent Persian Gulf War. One drug of choice now for seasickness or vertigo is the alkaloid scopolamine. Seasickness or airsickness can be a real morale downer during mass transport

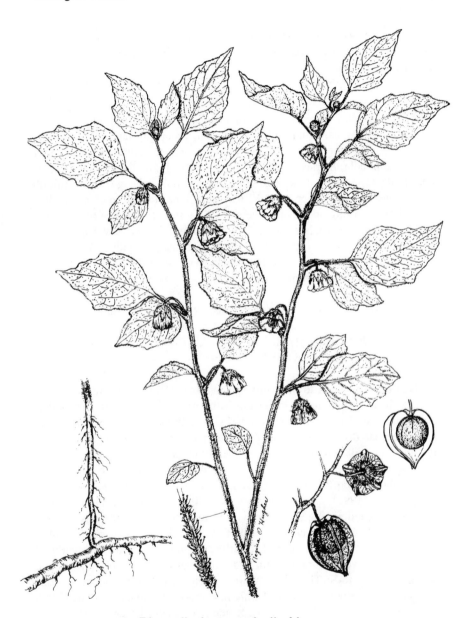

Physalis heterophylla Nees

of troops. The other strategic alkaloid from the potato family (SOLANA-CEAE) is atropine, the antidote for nerve gas. Most frontline troops in the Persian Gulf War had injectable atropine in special flaps in their pantlegs so that medics could automatically inject any fallen soldiers in case of the feared nerve-gas attack that never materialized in the war. WSSA lists 10 weedy species and/or varieties of *Physalis*.

PHYTOLACCA AMERICANA L.   (PHYTOLACCACEAE) --- Common Pokeweed, Pigeonberry, Poke, Pokesalad, Pokeweed

---

DESCRIPTION: Succulent perennial herbs to 9 feet tall, the stem often hollow, malodorous, often with a large, tan, branching taproot. LEAVES: one at the node, alternating on the stems, egg-shaped, usually broadest at or below the middle, hairless, shortly pointed at the tip, toothless on the edges, tapered to the short leafstalk, with several lateral veins arising from the midrib like plumes from a feather. FLOWERS: whitish or pinkish, in elongate clusters, often emerging on the opposite side of the stem from some of the upper leaves; sepals and petals similar, 5 in all (some scientists say 5 sepals, no petals); stamens 5–30, often 10; ovary 5- to 12-celled, with 5–12 terminal processes. FRUIT: a juicy, purplish-black, many-seeded berry, broader than long, about 0.2–0.4 inch in diameter.

DISTRIBUTION: A common and persistent weed, in full sunlight to dense shade, in old pastures, meadows, roadsides, woodlands, and waste places, flowering in July in Maryland, as early as May farther south, continuing to flower and fruit until frost. NE-ME-TX-FL. Zones 4–10.

UTILITY: I planned to omit this species, but I feel more foragers would criticize its omission than its inclusion, with caveat. To me, the fresh shootlets constitute one of the most tasty potherbs, readily harvested in spring. After frost, one can dig the poisonous roots and move them indoors. After a few months they will sprout, if kept sufficiently moist. If kept in a dark cellar, these sprouts may be almost white. Delicious! I am reluctant to pass on the stories about using the fruits for making candies, frostings, and pies (e.g., see Facciola, 1990). I had two visits in 1990 from poison control centers, because of children ingesting these attractive, but not salubrious berries. Moerman (1986) mentions dozens of folk medicinal uses, e.g., Cherokee took the berries or a wine thereof for arthritis and rheumatism; Mahuna used the roots for severe neuralgic pain and the leaves to remove blackheads and pimples. CAUTION: Though providing a delicious "mess" of greens, this herb can be extremely dangerous! Only new shoots should be consumed, with no woody or reddish tissue included, and even then, the greens should be boiled through two changes of water, discarding the pot liquor. There are reports of dozens of sicknesses among campers who followed all these precautions. Some well-respected scientists advise against even handling the plants, because of their mitogenic properties. Even though the berries were once used to tint (or taint) port wine, and even though the seeds have been ingested for arthritis, such practices may be dangerous. One lass learned the hard way, painting her body with the fruit juices for Halloween. She started itching violently shortly thereafter. But there's a flip side. Pokeweed contains

Phytolacca americana L.

three dangerous antiviral proteins, known technically as ribosome-inactivating proteins. With modern biotechnology, poisons such as these and the poisonous principle ricin of the castor bean can be tacked on, like hitchhikers, to mono-clonal antibodies, and directed selectively towards such enemies as the AIDS virus and tumors. And these compounds can affect the immune system, as well.

PLANTAGO MAJOR L. (PLANTAGINACEAE) ---
Broadleaf Plantain, Plantain

DESCRIPTION: Low perennial herb, stemless except for the stalk of the flower clusters to 20 inches tall, the roots fibrous. LEAVES: in a basal rosette, egg-shaped, 2–6 inches long, 1–4 inches broad, the tip pointed or rounded, the edges toothless, but sometimes wavy, tapered to the leafstalk, which may be shorter or longer than the blade, both occasionally with small hairs; the blade has about 7 veins arising near the leafstalk and extending nearly to the tip of the leaf. FLOWERS: greenish, inconspicuous, stalkless on the long central axis, the stalk of the clublike cluster usually longer than the leafstalks; sepals 4, united at the base; petals 4, fused a little higher; stamens 4; ovary 2-celled, free of the sepals, with 2 terminal processes. FRUIT: a small, 6- to 18-seeded pod, the top falling off like a cap (circumscissile); seeds reddish brown.

DISTRIBUTION: Weed, once called the "white man's footprint" by the Indians (reportedly introduced by the Puritans), in lawns, fields, roadsides, sand dunes, wherever white man has trod, flowering in June and fruiting on until frost. WA-ME-CA-FL. Zones 3–8.

UTILITY: Except for its abundance and ease of recognition, I would have excluded this potherb, mainly because I don't like it. But then, I don't like spinach, either. I've sampled all the plantain species I encountered around home and enjoyed none. Still, I would prefer plantain to starvation. Indians in New Mexico even learned to use the young leaves as a potherb. Mitich even found a recipe for sweet-and-sour plantain. One widely advertised "natural" laxative depends on the husks of the flower clusters of an Asian *Plantago* (also found in Carolina) known as "psyllium". I have stripped off the husks of our plantain to serve as a poor man's bran flakes or "psyllium". When I added the husks to my milk and sugar, I got the distinct impression that the husks were serving as rennet, curdling and/or souring the milk. You can get enough fiber eating the boiled young plantain leaves. Some books even suggest chopping the young leaves into salads. Facciola (1990) recommends selecting tenderer leaves, discarding the leafstalks, dipping in batter, and frying over low heat 30 minutes. Seeds are eaten parched or ground into meal (Facciola). Roots are also said to be edible. Some medical personnel are developing allergies to psyllium. Certain American firms wanted psyllium to be a food, others wanted to keep it an OTC (over-the-counter) laxative. Even Shakespeare mentioned plantain as a panacea. Its greatest fame as a medicine stems from the use of the leaves as a poultice, for everything from bee stings, to cancers, to cuts. Under the name "llanten" in Latin America, it appears in almost every herbal medicine market, with a well-deserved reputation as a

Plantago major L.

poultice on cancers. Leaves do contain allantoin, famous for alleviating skin sores. One 11th century leech book (early folk remedies) suggested the application of the cool leaves to travelers' sore feet, perhaps triggering the trend to call it the white man's foot or footprint. Its seeds long served as food, having been found in the stomachs of mummified ''bog people'' of 4th century northern Europe (Mitich, WT1:250.1987). WSSA lists 9 *Plantago* as weeds.

PODOPHYLLUM PELTATUM L.   (PODOPHYLLACEAE)
--- Mayapple

DESCRIPTION: Hairless, perennial, umbrellalike, malodorous herbs 12–18 inches tall, often forming "fairy rings" from branching, whitish, horizontal rootstocks. LEAVES: 1 or 2, arising from the simple or forked stem attached to the rootstock, round, 2–10 inches broad, umbrellalike, deeply lobed, the lobes sometimes with teeth, the leafstalk attached near the middle of the undersurface of the leaf. FLOWER: one, borne on a short stalk in the fork of the stem of the 2-leaved specimens, nodding; sepals 6; petals 6–9, white, longer than the sepals; stamens 12–18; ovary 1-celled, free of the sepals. FRUIT: a greenish, egg-shaped, many-seeded berry 1–2 inches long, 0.5–1 inches broad, turning yellow or brownish yellow when ripe, with an aromatic, edible pulp.

DISTRIBUTION: Deep shade of deciduous forests, sometimes rather common in alluvial situations, or persisting in pastured forests, flowering March to April in Carolina, April to May in Maryland, the fruits ripening late July to August, often after the leaves have fallen off or been grazed by deer, such that there is a naked forked stem with the mayapple fruit in the fork. MN-ME-TX-FL. Zones 3–7.

UTILITY: The fruits, when ripe, provide a pleasant change, with flavor and aroma not exactly like anything I know, blending the mysteries of citrus and passionfruit, with a hint of quince. Eating pulp of one fruit with the seeds intact taught me that the seeds, like those of tomato, wild cherry, and grape, pass through largely undigested. That's probably good, because the seeds probably contain toxins. The fruit pulp can be used to make mayapple sauce, mayapple madeira, and mayapple lemonade, among other things. I've even enjoyed mayapple ice cream. The taste gets overwhelming after awhile. One year, we made 8 pounds of mayapple sauce, some of which stayed in the freezer a long time. Some was even given, on request, to a testicular-cancer patient. Like other fruits, mayapple has good and bad years for fruit production. I got almost no fruit set in the 1989 and 1990 seasons. A lot of people, including me, have gone too far with the claim that Penobscot Indians used the mayapple for cancer. It cannot be convincingly documented in the literature, and there are few if any mayapples in current Penobscot country in Maine. CAUTION: All parts of the plant except the fruit are poisonous, but are also sources of commercial medicines, very important in venereal warts and cancer. Etoposide, a semisynthetic compound derived by modifying the podophyllotoxin extracted from mayapple rhizomes and roots, is now commercially available for testicular cancer and small-cell lung cancer. The Indians taught us the smoking habit, lobelia to curb the smoking habit, and

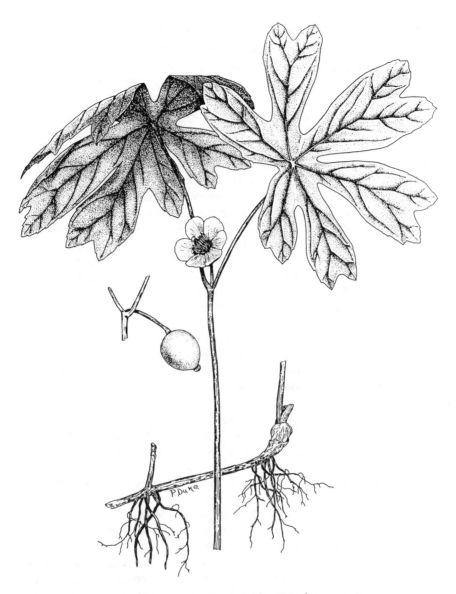

Podophyllum peltatum L.

finally mayapple to help some of the lung cancers generated by smoking. I suspect the seeds, if chewed, would be bitter and sickening. But I will chew one seed, dangerously, of the next mayapple I see, perhaps dosing myself homeopathically with podophyllotoxin. Homeopaths believe that minute doses of a compound may cure or prevent what large doses cause. WSSA does not list this as a weed, although it does persist as a weed in forested pasture.

POLYGONUM CUSPIDATUM Siebold and Zucc. (POLYGONACEAE) --- Japanese Bamboo, Japanese Knotweed, Mexican Bamboo

DESCRIPTION: Erect perennial herbs to 10 feet tall, with a waxy, gray, hollow stem and a thick rootstock. LEAVES: one at the node, alternating on the stem, egg-shaped to nearly heart-shaped, 3–6 inches long, 2–5 inches wide, broadest below the middle, shortly pointed at the tip, toothless on the edges, basally squared off at the leafstalk, the midrib with several lateral nerves arising like plumes from a feather, the leafstalk swollen at the stem, with a small, cufflike flange (ochrea) embracing the stem. FLOWERS: green-ish-white, the males on one plant, females on another, in long, branching clusters 3–6 inches long in the angle between the stem and upper leaves; sepals 4–5, somewhat petallike; petals 0; stamens 3–9, often about 8; ovary 2- to 3-celled, with 2–3 styles. FRUIT: a small nutlet, 3-sided in cross section, closely embraced by the enlarged sepals, the outer sepals flaring out like a wing along the midrib; nutlet glossy black.

DISTRIBUTION: Rather recently introduced weed, here and there forming large stands, in neglected lawns and gardens, alluvial floodplains and forest borders, flowering in May in Carolina, midsummer in Maryland, the fruits ripening near frost. MN-ME-TN-NC, Zones 4–7.

UTILITY: The sprouts, harvested (March to April in Maryland) and cooked like asparagus, are this herb's primary culinary contribution to the forager's larder. Sprouts up to a foot tall are boiled 3–5 minutes. Even leafy tips of shoots up to 2 feet tall may be cut and boiled, producing a dish suggestive of the related French sorrel. Foragers with large stands of the herb may stimulate fresh shoots in summer by cutting back the adults. Tender parts of the rootstock are also peeled, boiled, and buttered as a potatolike dish. "Rhu-barb sauce" may be simulated by peeling the young stems and cooking them slowly in sugar water. Containing the antiseptic and antitumor compound emodin, the plant is viewed as a prophylactic in epidemics in China. Chinese also use it for abscess, appendicitis, arthritis, boils, bruises, burns, dysme-norrhea, gout, hepatitis, and traumatic injuries (Duke and Ayensu, 1985). CAUTION: Like rhubarb, this may be laxative. Novices should sample ju-diciously, if at all. WSSA lists 22 weedy *Polygonum*. Among these we find several also listed as edible by Facciola (1990), Tanaka (1976), or Yanovsky (1936): *P. amphibium,* the Water Knotweed; *P. aviculare,* the Prostrate Knot-weed; *P. bungeanum,* the Prickly Smartweed; *P. douglasii,* Douglas Knot-weed; *P. hydropiper,* the Marshpepper Smartweed; *P. lapathifolium,* the Pale Smartweed; *P. orientale,* the Princess-Feather; *P. persicaria,* the Ladys-thumb; *P. punctatum,* the Dotted Smartweed; *P. sachalinense,* the Sakhalin

Polygonum cuspidatum Siebold & Zucc.

Knotweed; WSSA has not gotten around to listing our worst weed in the genus *Polygonum, P. perfoliatum,* the ''perfoliate smartweed'', an aggravating, blue-berried vine taking over some streamsides haunted by foragers. Treating it under the name *Chylocalyx perfoliatus,* Tanaka (1990) states that the young leaves, stem tops, and young fruits are cooked with other vegetables, or used in stews as a tamarind substitute, or added to fish dishes. I predict that the Chinese *P. multiflorum* will join the ranks of U.S. weeds next century.

PORTULACA OLERACEA L. (PORTULACACEAE) ---
Common Purslane

DESCRIPTION: Prostrate, succulent, annual herbs, the branching alternate or opposite, the stems often tinted with red. LEAVES: one at the node (alternate) or rarely two on opposite sides of the node towards the tips, 1–2 inches long, 0.3–0.8 inch broad, spoon-shaped in outline, rounded at the tip, toothless and hairless at the margin, basally tapered to the stem, essentially stalkless, the midrib evident, other veins obscure. FLOWERS: yellow, almost stalkless in the angle of the stem with the leaf; sepals 2, basally united into a tube; petals 5; stamens 6–12; ovary 1-celled, embraced by the sepal tube, usually with 3 styles. FRUIT: a small pod, the top coming off like that of a jack-o-lantern, 0.2–0.4 inch long, longer than broad, with many reddish to black seeds.

DISTRIBUTION: Common weed in gardens, fields, large unkempt lawns, waste places, etc., usually in full sun, flowering May to June, seeding shortly thereafter. WA-ME-CA-FL. Zones 3–10.

UTILITY: Early available as a potherb or a salad ingredient, and later available as a cereal, this mucilaginous potherb has been used by many ethnic groups on many continents. ''Only in the United States has fame and favor eluded this hot-weather weed'' (Jones, 1991). Tender parts, leaves, flowers, pods, seeds, and stems all may be stewed as potherbs, improved by adding egg and/or breadcrumbs. To me, the potherb tastes like a slimy spinach. Purslane recently received undeservedly good press as a vegetarian source of omega-3 fatty acids (with many of the health attributes attributed to fish oils). Actually, seed oils from the cabbage and walnut families are much better sources. USDA analyses (Norman, 1991) indicate this may be extremely high in alpha-tocopherol (vitamin E), making it a nutritional double whammy. The succulent stems may be pickled in brine or pickle vinegar. Facciola (1990) suggests another way to preserve this copious summer food for winter: put it into salt and dry white wine. The seeds are gathered, rather tediously I'm sure, and used for making flour or porridge. I suppose it's safe to eat sprouts of any plant like purslane, where the seeds and the plant are considered edible. But one way to learn the exception is to postulate a rule. Ashes of burned purslane serve as a good salt substitute. Buntings, doves, larks, longspurs, mice, and sparrows eat the seed. Rats and rabbits graze the foliage. Weed scientists tend to overlook its value as food, instead stating it is now ''used widely as a food for pigs'' (Holm et al., 1977). So there may be a bit of purslane in a pig's ear. Greeks are feeding their chickens purslane to increase the levels of omega-3 fatty acids. The levels apparently are increased, but tend to be unstable. Cherokee squeezed purslane juice into the ear for earache, also used the plant for worms. Iroquois poulticed it onto bruises and burns.

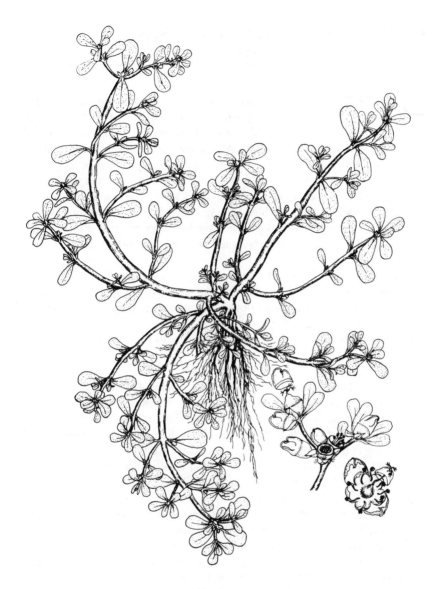

Portulaca oleracea L.

Navajo used it for pain and stomachache, almost as a panacea (Moerman, 1986). Chinese use it for anthrax, boils, bug bites, colic, dermatitis, eczema, enteritis, erysipelas, herpes, indigestion, leukorrhea, ophthalmia, piles, snake-bite, and tumors (Duke and Ayensu, 1985). Listed as a weed by WSSA, and listed by Holm et al. in *THE WORLD'S WORST WEEDS* (1977) with several other edibles among the worst 18 weeds of the world, more than half of which are edible and hence treated in this book on edible weeds.

PRUNELLA VULGARIS L. (LAMIACEAE) --- Carpenter Weed, Eel Oil, Healall, Selfheal

DESCRIPTION: Green, clambering or erect, perennial, square-stemmed, herbaceous mint sometimes nearly 3 feet tall. LEAVES: egg-shaped or more elongate, stalked on opposite sides of the stem, with 3–7 veins coming off the middle vein and arching out towards the margins, tapered to the leafstalk, pointed at the tip, the blades 1–4 inches long, usually 1 inch or less wide, the stalks short or up to 2 inches long. FLOWERS: numerous and small, very short-stalked, the clusters topping off the uppermost branch or branches, with 5 greenish sepals around the blue to purple snapdragonlike flowers, with 4 pollen-producing stamens tucked inside the flowers surrounding the nutlets. FRUIT: 4 minute seedlike nutlets which function as brownish seeds.

DISTRIBUTION: Barnyards, fields, grasslands, meadows, old homesites, powerlines, roadsides, volunteering in unkempt lawns, or adapting to regular mowing in lawns with a dwarf flowering variety; seeming to thrive in full sun to dense shade. Starting to flower in April in the Carolinas, May in Maryland, fruiting up to frost. Not so sensitive to frost as basil. WA-ME-CA-FL. Zones 3–9.

UTILITY: 'Twas a Labor-Day weekend in Pendletown County, WV, with me and the *Poa* boys playing bluegrass music way back in the hills, when I learned that my hosts ate this plant, under the interesting name "eel-oil", a corruption of the name "heal all". They boiled it as a potherb up there. Since it is proving to have some antiviral activity, I look forward to my first "mess" of "eel-oil greens". If it tastes bad, float it in vinegar, hot sauce, and onions, and surround it with cornbread. Facciola (1990) adds that young shoots and leaves are eaten raw in salads, cooked with other greens as a potherb, or added to soups and stews. He suggests a refreshing beverage made by soaking the leaves, freshly chopped or dried and powdered, in cold water. British Columbia Indians use the cold tea as a beverage, perhaps boosting their immune systems in the process. The infusion was taken for any ailment, such that both Indians and Chinese regard it as a panacea, while the Caucasians seem to have mostly lost interest therein. Bella Koola took it for heart conditions; Cherokee for acne, bruises, burns, cuts, and diabetic sores; Chippewa took the roots in polyherbal laxatives; Cree chewed it for sore throat (perhaps reflecting its antiviral properties); Delaware used it for fever; Iroquois for backache, biliousness, cold, cough, cramps, diarrhea, fever, shortness of breath, sore legs, stiff knees, and the like; Menominee for dysentery; Mohegan for fever; Ojibwa for female ailments; Quileute and Quinault rubbed the plant on boils; and finally, the Thompson Indians took the tea, hot or cold, as a tonic for general maladies or indisposition (Moerman, 1986). Chinese took

Prunella vulgaris L.

the seed for anxiety, headache, high blood pressure, hepatitis, ophthalmia, scrofula, and tinnitus, and the leaves and/or flowering shoots for boils, cancer, conjunctivitis, gout, fever, hepatitis, rheumatism, scrofula, and tumors. Recent Chinese studies indicate immune-boosting potential, and it is being studied for anti-AIDS potential. The recent discovery that *Prunella* is a best source of antioxidant rosmarinic acid may rekindle Occidental interest.

PRUNUS SEROTINA Ehrhart   (ROSACEAE) --- Black Cherry, Wild Cherry

DESCRIPTION: Large deciduous tree to as much as 100 feet tall, the inner bark aromatic, the outer bark dark gray to nearly black, the twigs marked with long lines perpendicular to the long axis. LEAVES: one at the node, alternating on the stem, willowlike to narrowly egg-shaped, 2–6 inches long, 1–2 inches wide, apically drawn out into a point, marginally saw-toothed, basally drawn out on the short leafstalk, the midrib with several strong nerves leading out towards the edges. FLOWERS: white, in elongate clusters 2–6 inches long at the tips of leaf branches; sepals 5; petals 5; stamens numerous; ovary 1-celled, free of the sepals, terminating in a single process. FRUIT: a 1-seeded black or purple berry.

DISTRIBUTION: Frequent weedy tree in fencerows, thickets, and woodlands, and seemingly tolerating the shade of deciduous forests, more rarely coniferous forests, flowering in April and May, the first fruits ripening in July and sometimes continuing on until frost, more often quickly devoured by birds. ND-ME-AZ-FL. Zones 4–9.

UTILITY: Unlike most of the authors, I enjoy the fruits out of hand, eating as many as 50 on a jogging lunch hour, sometimes even swallowing the seeds whole or masticated, suspecting that they might contain traces of laetrilelike, but dangerous compounds. I also add the cherry juice to lemonades or to color homemade liqueurs. Copious as the fruits are in late July and August, they can be used for pies, jellies, etc. Indians pounded up wild cherries, seeds and all, leaching the poisons out of the seeds. They then dried them in the sun, for "reconstitution" later in the leaner times of winter, or added the paste to pemmican or stews. Facciola (1990) adds that the slightly bitter fruits are eaten fresh, stewed, or made into "cherry bounce", jellies, pies, and sherbets, and for flavoring brandy, cider, rum, and other liqueurs. Though Indians were said not to have known alcohol before the arrival of the white man, Menominee reportedly became intoxicated eating black cherries that had been picked and allowed to stand for some time. The bark extract is used in flavoring baked goods, candies, soft drinks, and syrups (Facciola, 1990) and shows up in many over-the-counter (OTC) cold and cough preparations, especially in combination with white pine (Leung, 1980). Numerous frugivorous birds feed on the fruits, mammals harvesting what the birds drop to the ground. Browsing animals eat the bark and twigs, sometimes loaded enough with cyanide to poison domesticated animals. This species is the most dangerous of the eastern wild cherries. Laetrile is not among the 46 cancer-preventive compounds listed by Stitt (1990) for the tame cherry. CAUTION: Four ounces of foliage could kill a 100-pound animal. Some sources claim

Prunus serotina Ehrhart

that children have died after eating the seeds. Fall-collected bark may have 1500 ppm HCN, spring-collected bark only 500 ppm; leaves may contain as much as 2500 (Leung, 1980). Cyanide poisoning has been reported among neo-Amerindians from improperly prepared "pemmican" (Wagstaff, FDA; pers. comm.). Strangely, WSSA did not list *P. serotina,* the most common and one of the more edible species of *Prunus,* as a weed, while they did list 11 species and/or varieties, including cherries, peaches, and plums as weeds.

PUERARIA LOBATA (Willd.) Ohwi   (FABACEAE) ---
Kudzu

DESCRIPTION: High-climbing herbaceous or woody deciduous vine, sometimes 100 feet long, the young stems densely hairy, with swollen, starchy roots. LEAVES: one at the node, alternating on the stem, composed of three leaflets, these narrowly to broadly egg-shaped, 2–8 inches long, apically pointed, marginally toothless to lobed like an oak, basally rounded or notched for the attachment of the short leaflet-stalk, the leafstalk usually longer than the leaflets, hairy. FLOWERS: purple, with the aroma of artificial grape soda, in long clusters emerging from the angle where the leafstalk meets the stem, the clusters 4–12 inches long, hairy; sepals 5, basally united; petals 5, in a pea-shaped arrangement; stamens 10; ovary 1-celled. FRUIT: an elongate, beanlike pod, hairy, many-seeded, 1–2 inches long.

DISTRIBUTION: Old homesites, roadsides, fields, and deciduous or con- iferous forests, sometimes climbing and weakening tree species in the forests, starting to flower in July or August, fruiting up until frost. MO-NY-TX-FL. Zones 5–9.

UTILITY: Several authors, who probably have not prepared much flour themselves, speak of the virtues of the flour, derived by peeling, pounding, and soaking the starchy parts of the tuberous swellings on the root. Orientals like the starch because it doesn't become as chewy and elastic as corn and potato starch, and because it tastes better and gives a smooth, translucent appearance to desserts (Dharmananda, 1988). Elliott's first-hand account (1976) describes more than a day's work, netting him half a cup of "watery sediment ... certainly not worth the work". My experiments were no more rewarding. I did enjoy some fried kudzu leaves dipped in batter. They had the marvelous taste of fried batter. I suspect most of the leaves discussed in this small volume would turn out as well as fritters, if well battered. I hope that someone, incensed by my negative kudzu report, will help us all by wiping out one more plant to repeat Elliott's experiments. There are many people who think more highly of the kudzu as food. You can buy several ounces of pure kudzu flour at your Japanese market for considerably less than a day's wages. Some people cook diced roots in their soups as thickeners, not necessarily eating the kudzu "cubes" (Rogers and Powers-Rogers, 1988). Facciola (1990) even hints that the roots are boiled or steamed and served with miso, salt, or soy sauce. He adds that young leaves and shoots are eaten raw, or boiled, fried, pickled, or sauteed. Pickled shoots are rather intermediate betwixt beans and peas in taste. The grape-scented flowers are cooked or pickled and eaten. One of our most despised weeds, kudzu can produce up to 8 tons/acre biomass which might be used for energy generation or fodder. Dharmandanda (1988) notes that in 1 century, 1 acre will expand to cover 13,000 acres. So recently

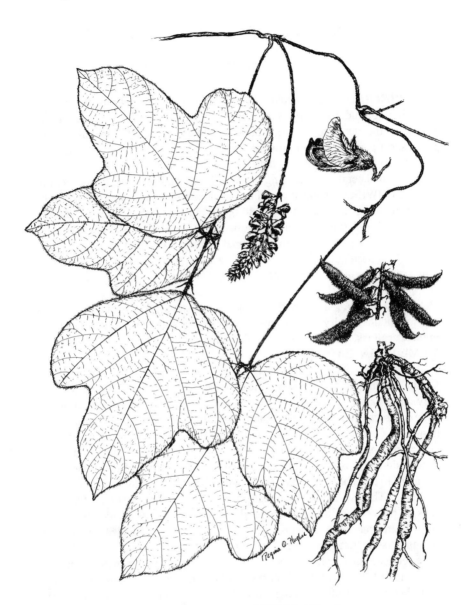

Pueraria lobata (Willd.) Ohwi

arrived in America, the kudzu has not yet moved into Amerindian folk medicine (Moerman, 1986), but it is increasingly important in oriental medicine. Flavonoids, constituting perhaps 5–10% of peeled roots (Dharmananda, 1988), appear to lie behind the medicinal virtues of the plant. At a 1990 NCI soybean symposium, flavonoids like daidzein were touted as cancer preventives. Perhaps wild legumes might contain even more of these compounds.

QUERCUS PRINUS L.   (FAGACEAE) --- Rock Chestnut Oak

DESCRIPTION: Large deciduous trees to nearly 100 feet tall, the tannin-laden bark on older trees deeply furrowed. LEAVES: one at the node, alternating on the stem, egg-shaped, broadest above the middle, apically drawn into a short point, with deep, rounded teeth or shallow lobes along the margin, tapering to the leafstalk, the blade 4–8 inches long, 2–4 inches wide, with conspicuous lateral veins leaving the midrib for the tips of the major teeth or lobes. FLOWERS: greenish or yellowish, inconspicuous, appearing with the leaves in spring, the male in long, drooping clusters with 2–8 sepals, basally united, and 3–8 stamens; females solitary or clustered in the angle formed by the leafstalk with the stem, the sepals 6, basally united; ovary 3-celled, ending in 3 terminal processes. FRUIT: a large acorn, brown, an inch or so long, nearly an inch in diameter.

DISTRIBUTION: Trees of the deciduous forests, especially of the piedmont and mountains, in dry, sandy, gravelly, or rocky soil, flowering about April in the Carolinas, May in Maryland, the fruits maturing in the fall about frost. MI-ME-IL-GA. Zones 4–7.

UTILITY: I selected this from the dozens of oak species because it is the only one I have found that had, on rare occasions, acorns that were fit to eat without processing. I have also eaten sprouting seeds in fall and spring, and find some of them edible as is. Many of the oaks have very bitter acorns that can be rendered edible, if not palatable, by prolonged leaching in alkaline water (add wood ashes to the water to achieve this alkaline effect). One might try a trick the Andean Indians use with their lupines to leach out bitter alkaloids. They stick the bitter seed in a sock in a running stream for a few days. In many cases, the resulting product is not worth the considerable effort, but with the chestnut oak acorns, the story sometimes has a pleasant culinary ending. Astute foragers have come to recognize out east that they have a better chance for a palatable acorn on broad-leaved species without bristles at the tips of the leaf lobes. Acorns can be candied, or ground into grits or flour that may be well worth the effort. Flour derived from the sweeter acorns can be used to make breads and muffins, but I have never made these myself. I have made johnnycakes out of fried acorn patties. Scorched acorns have served as coffee substitutes. Acorns were almost staple foods with some Indians, but they spent a lot of time processing them. Though I regard chestnut oak every bit as weedy as the White Oak (*Quercus alba*), WSSA lists the latter among their 19 weedy oaks. Facciola (1990) suggests leaching and baking white oak acorns, especially the "sweeter" (=less astringent) ones, as indicated by pink or red blotches on the stem, in butter and salt, "until

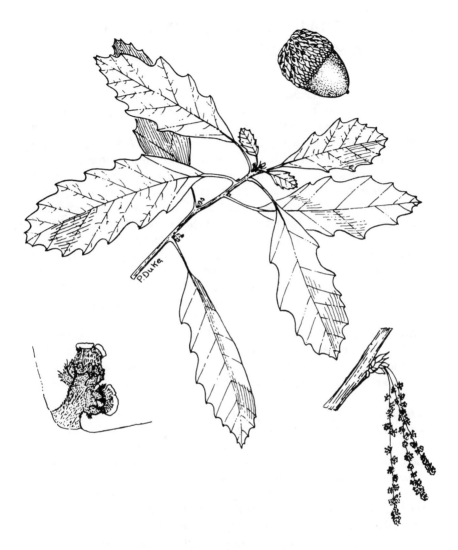

Quercus prinus L.

the acorns no longer taste raw''. Cooled roasted acorns are compared to something between sunflowers and popcorn. Chestnut oak seedlings often cover the forest floor in middle Maryland. CAUTION: While foragers will unlikely experience problems with minor acorn indulgence, cattle pastured in oak groves may suffer ''acorn poisoning'' (Kingsbury, 1964), becoming ''hooked'' on acorns, avidly seeking them out to the exclusion of other foods. Animals grazing exclusively on oak leaves or acorns may have big problems.

RHUS GLABRA L. (ANACARDIACEAE) --- Common Sumac, Smooth Sumac

DESCRIPTION: Handsome deciduous shrub or small tree, often forming thickets, 15–25 feet tall, with thickened rootstocks, the twigs smooth and 3-sided (not hairy and round as in the staghorn sumac). LEAVES: one at the node, alternating on the stem, compound, pinnate, with 7–9 (to 15 in staghorn sumac) pairs of leaflets arising like plumes from a feather, the midrib, and a similar terminal leaflet; leaflets stalkless, narrowly egg-shaped, or broadly willow-shaped, curved a bit and pointed at the tip, saw-toothed on the margin, rounded at the base, 2–7 inches long, 1/2–2 inches wide, with several strong lateral nerves arising at the midrib and extending towards the margins. FLOW-ERS: in long clusters at the tips of the branches, greenish or greenish yellow, sometimes with the males on separate plants; sepals 4–5; petals 4–5; stamens 4–5; ovary 3-celled, ending in 3 terminal processes. FRUITS: green, then pink, then maroon to red, in large, tight terminal clusters to 10 inches long, 4 inches broad, the individual berries small, hairy, each with 1 large seed.

DISTRIBUTION: Borders of woodlands, roadsides, pastures, thickets, and waste ground, either dry or moist, but open situations, flowering in May in the Carolinas or June to July in Maryland, fruits ripening as early as June and persisting into the winter as a nibble. WA-ME-CA-FL. Zones 4–9.

UTILITY: Perhaps this belongs more in a beverage than a food book, but when jogging and too thirsty, I have found that eating these fruits, even chewing up the seed, tends to quench the thirst, leaving a pleasant taste in the mouth. I have not experienced three consecutive nights of flying dreams (though I have them intermittently) as Millspaugh (1892) reports following prolonged contact with the sumac. Next summer I'll try sleeping on sumac boughs. On camping trips, I have added the reddish fruits, better harvested before than after rain, to cold or boiling water to make a poor man's lemonade, or to sassafras tea as a lemon substitute. Cool water seems to extract less of the astringent tannin than hot water extracts. I find that sumac-sassafras tea is much better than the sum of its parts. Before the fruits have ripened sufficiently (early June in Maryland), I have eaten them as a forager's cotton candy. Sumac extract, tart as it is, makes a pleasant natural additive to other fruit juices, for example, black cherry, elderberry, or wild grape, in several possible combinations. With sugar and pectin, such pleasant beverages can be converted to jellies. Indians stored the fruits overwinter, but around Maryland at least, Father Nature stores them rather well on the bush. Indians peeled fresh roots (Yanovsky) and ate them raw. Fenton, on the other hand, recalls that Iroquois ate emerging sprouts in spring. Jones (1991) advises us that

Rhus glabra L.

Indians smoked dried sumac berries, serving as an alternative to tobacco. CAUTION: Some people hypersensitive to poison oak and ivy ("leaves of 3") and poison sumac (swamp species with white berries) may be allergic to the so-called safe sumacs. WSSA lists 4 weedy species of *Rhus,* assigning the Poison Ivy, Poison Oak, and Poison Sumac to the genus *Toxicodendron.*

ROSA MULTIFLORA Thunb.   (ROSACEAE) ---
Multiflora Rose

DESCRIPTION: Erect, thorny, green-stemmed shrub to as much as 20 feet tall, occasionally forming impenetrable thickets. LEAVES: one at the node, alternating on the stem, pinnate, i.e., with 3–4 pairs of opposed leaflets on a midrib terminating in a similar leaflet; leaflets egg-shaped, 1/2–2 1/2 inches long, broadest below, at, or rarely above the middle, apically drawn out into a point, marginally saw-toothed, basally rounded to a short stalk, the leafstalk hairy. FLOWERS: white, in few- to many-flowered clusters terminating the branches; sepals 5; petals 5; stamens numerous; ovules numerous, finally enclosed within the sepals. FRUIT: a reddish, juicy, egg-shaped berry, with the tips of the sepals protruding at the tip of the fruit, 1/4–1/3 inch long.

DISTRIBUTION: A serious weed, but appreciated by some (like my mother, who calls it tea rose), taking over fencerows, waste places, openings in the deciduous forests, poorly tended lawns, and pastures, etc., flowering May to June, the fruits ripening shortly thereafter, hanging on late into the winter. It seems to battle well with the honeysuckle and kudzu, other introduced vines in Maryland. WA-ME-CA-AL. Zones 4–8.

UTILITY: Fortunately for hikers, there is not much to confuse with this copious producer of nutritious but seedy rose hips. Around Herbal Vineyard, one can indulge in the fruits well into February, but by then the red color has blackened a bit. The fruits, like the petals, can be added to teas, salads, fruit cups, jellies, and the like. I was surprised at how astringent the raw petals are. Still, I have consumed many of the petals and fruits during a noon run. Indians ate the buds and young shoots of other species. Facciola says budlings and young leaves are parboiled and eaten. Rose petals are finding their way into many floral foods in the current flower fad; in a recent paper, *TAKE TIME TO EAT THE ROSES* (1990), I compared rose flowers unfavorably with several available to me in October at Beltsville:

Generic name	Common	Taste	No. flowers I Consumed	Rank
Cichorium	Chicory	Astringent	Dozens	6
Cirsium	Thistle	Chewy	Hundreds	3
Malva	Mallow	Slimy	Twenty	5
Oenothera	Evening Primrose	Tasteless	Ten	4
Oxalis	Wood Sorrel	Tart	Thirty	1
Polygonum	Tearthumb	Astringent	Dozens	9
Rosa	Rose	Bland; Astringent	One	7
Sonchus	Sow Thistle	Chewy	Hundreds	2
Taraxacum	Dandelion	Chewy; Astringent	Hundreds	8
Trifolium	Red Clover	Leguminous	Dozens	10

Rosa multiflora Thunb.

Even the birds seem to tire of the fruits; but when snows have hidden other fruits, the birds may once again seek out the rose hips. A wealth of folk medicinal uses are reported by Moerman (1986) and Duke (1986). WSSA lists 10 more weedy roses, all apparently with edible ''hips'', rich in vitamins and minerals and freely available.

RUBUS SPP. (ROSACEAE) --- Blackberry, Dewberry, Raspberry

DESCRIPTION: Arching, trailing, or erect, usually thorny, often gregarious, thicket-forming perennials. LEAVES: in the edible species 3- or 5-fingered (i.e., compound, with 3 or 5 leaflets), one at the node, alternating on the stem, the leaflets 1–5 inches long, egg-shaped, broadest below, at, or above the middle, apically pointed, marginally saw-toothed, basally tapered or rounded, with lateral nerves arising from the midrib like plumes on a feather, the leafstalks shorter, equalling or longer than the leaflets in length. FLOWERS: white, rarely pinkish, solitary or in clusters terminating the lateral branches; sepals 5, hairy; petals 5; stamens many; ovaries many and at first separate. FRUIT: a berry composed of the fused ovaries, black in blackberries and dewberries and some raspberries, red, yellow, or purple in raspberries, and red in wineberries.

DISTRIBUTION: Old gardens, hedgerows, fencerows, and in openings in the coniferous and deciduous forests, flowering from April to August, and fruiting from June to frost. The genus *Rubus* as a whole offers us some fruit all summer, though a given species may have a short fruiting span. Around Herbal Vineyard, the red raspberry is first to ripen (some by June 1), closely followed by the purple raspberry and wineberry, then by wild blackberries. A second, but smaller crop of red raspberries often accompanies the fall grapes at the first frost. Though most were brought in as cultivars, all have made themselves at home and moved out of bounds, often quite aggressively. *Rubus allegheniensis:* MN-ME-MO-SC. Zones 3–7.

UTILITY: At Herbal Vineyard, the first day of June is usually enriched by the first of my raspberries and blueberries, with the last of my ''tame'' cherries and strawberries. Although I enjoy the blackberries and raspberries out of hand, I also add them to lemonades to make pink lemonades of superior flavor. Of course, there is no end of desserts and fruit cups and juices and wines that can be combined from the numerous varieties of the genus *Rubus*. Young shoots, peeled of their thorns, are edible, but often astringent. Sprouts of some species are eaten like rhubarb. Young shoots of *R. allegheniensis,* e.g., are used in salads. Leaves of all are suggested for pleasing yet medicinal teas, especially *apropos* for women once a month. I just as readily eat the smaller petals of the flowers of these brambles as I do the flowers of the roses, mentioned earlier. CAUTION: Still, the tea made from leaves of black raspberry, *R. occidentalis,* has been said to be physiologically harmful. WSSA lists 11 weeds, including <u>Blackberries, Dewberries, Raspberries</u>, and <u>Thimbleberries</u>. Around Herbal Vineyard and generally in Maryland, the wineberry, *R. phoenicolasius,* may be my worst weedy *Rubus*. It belongs on the

Rubus allegheniensis Porter

WSSA list. But the birds get most of its berries. Hence I've yet to indulge in wineberry wine. Brambles furnish just one of dozens of examples of *PARALLELS IN CHINESE AND AMERINDIAN PHYTOTHERAPY* (Duke and Ayensu, 1985). Although Amerindians did not practice acupuncture, they frequently used the thorns of brambles to inject other medicines superficially. And both Chinese and Amerindians burned *Artemisia* in traditional medicine. Both used the astringent species of *Rubus* as a hemostat.

RUMEX ACETOSELLA L.   (POLYGONACEAE) --- Red Sorrel, Sheep Sorrel, Sour Grass

DESCRIPTION: Low, weedy, perennial herbs, to as much as 1.5 feet tall, often with a basal rosette, the stem 4-angled, striate. LEAVES: of the rosette halberd-shaped, 2–6 inches long, long-stalked, the stalk basally with a cuff around the stem; leaves of the stem one at a node, alternating on the stem, hairless, smaller, apically barely pointed, marginally toothless, basally squared off or notched at the point of attachment of the long leafstalk, which is often longer than the blade; 3–5 veins arise at the junction of the leaf base and leafstalk. FLOWERS: greenish yellow to reddish, arising in the angles formed by the upper leaves with the stems, those on male plants small, with 6 sepals and 5 stamens; those of the female plants with 6 sepals and a simple ovary terminating in 3 processes, the ovary basally attached to the sepals. FRUIT: a small nutlet partially embraced by the persistent sepals.

DISTRIBUTION: A weed partial to poor acid soils in gardens, pastures, old fields, thriving in acid berry patches, usually in partial or full sun, occasional in sandy soils, flowering March to June and fruiting rather rapidly thereafter. In Maryland I seem to have a spring crop, as early as February, and a fall crop, although I can find scraggly crops in midsummer. WA-ME-CA-FL. Zones 3–8.

UTILITY: Sheep sorrel, loaded with oxalic acid, shares the name sorrel and sourgrass with members of the genus *Oxalis*, which also contain oxalic acid. I enjoy eating the leaves out of hand as a nibble, and have cooked up big batches of the acid greens. Once I got bold and drank all the pot liquor (about 4 cups) from a huge batch of sorrel greens I had prepared for some visiting dignitaries. Then I got worried about oxalic acid when I felt minor prickling sensations of the tongue. I told the nurse I had eaten too much rhubarb (of the same family), and she suggested milk as a source of an antidote, calcium. In a survival situation, I would have merely taken cool ashes from the campfire as a source of calcium. Sheep sorrel can be substituted for the rarer French sorrel in recipes for sorrel soup, boiled vigorously and mashed into a puree, seasoned with butter, rice water, and spices. Irishmen early learned to make soups of sorrel and milk. Scandinavians add it to bread (Jones, 1991). Stems, seeds, even the roots, of some *Rumex* species were eaten by Indians, probably in inverse proportion to the tannin content. CAUTION: Oxalic acid levels may attain 10–35% of dry matter in some species, most of which average about 90% water. The normal USDA serving of 100 grams fresh sorrel would be equivalent to 10 grams of dry sorrel, 10–35% of which could be oxalic acid. That translates to 1000–3500 milligrams of oxalic acid per 100-gram serving. The lowest lethal dose (LDlo) reported for

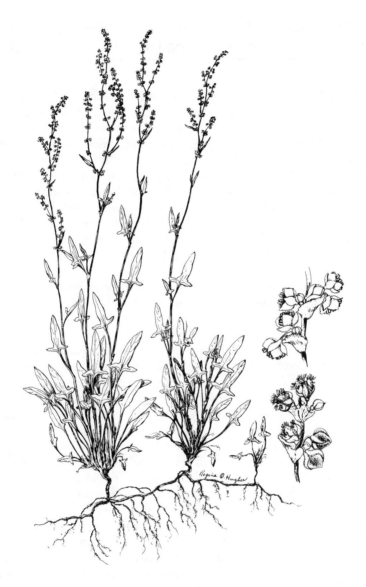

Rumex acetosella L.

humans is 700 mg/kg body weight. Since I weigh 110 kg, my LDlo would be 77,000 mg. If my half-cup serving (100 grams) had 1000 mg oxalic acid, it would take 77 servings to attain my LDlo; if it contained 35% oxalic acid, it would take closer to 20 servings (nearly 5 pounds of sorrel). Only once did I eat more than one serving. WSSA lists 13 weedy species in the genus *Rumex*.

SAGITTARIA LATIFOLIA Willd.   (ALISMACEAE) ---
Arrowleaf, <u>Common Arrowhead</u>, Duck Potato, Wapato

DESCRIPTION: Hairless, perennial, aquatic or marsh herb to as much as 5 feet tall, the horizontal roots bearing fleshy buds at their tips in autumn. LEAVES: all in a rosette, strap-shaped, egg-shaped, or most frequently arrow-shaped, the blades 2–16 inches long, the leafstalks even longer, the tips drawn into a short point, the margins toothless, the base arrowheadlike, with all the main veins arising at the juncture of the leafstalk and blade. FLOWERS: white, in long-stalked clusters, the individual flowers whorled around the elongate axis, upper flowers usually male, the lower flowers usually female; sepals 3; petals 3; stamens 15–40; "ovaries" simple, numerous in the rounded centers of the female flowers. FRUITS: in rounded green clusters of greenish, one-seeded nutlets, the clusters to 1 inch wide.

DISTRIBUTION: Often in sluggish streams, pond margins, or open places in marshes and swamps, flowering in June to July, then flowering and/or fruiting on up to frost. WA-ME-CA-FL. Zones 3–8.

UTILITY: How many times I read the story! Indian squaws waded out into the mud in the cold autumn waters, dislodging the wapatoes with their colder toes. So I decided to go through the ritual for Neil Soderstrom's camera. I like the photo, on the dust cover of my *Handbook of Northeastern Indian Medicinal Plants* (Duke, 1986). With my long arms I could grope through the mud, dislodging the wapatoes with my fingers. The wapatoes, shaped like germinating onion sprouts, then floated to the surface where I gathered them, eating only one raw, saving the others for analysis or cooking. They are a bit starchy raw, much improved by boiling and buttering. Indians would roast them in ashes or slice them, drying them in the sun for winter use. They also cooked them with venison and maple sugar, and candied them in maple sap. Chinese immigrants, new in America, adopted this food species, in lieu of the Asiatic species they had enjoyed in Asia. The tubers are often connected to the mother plant by long roots. These too have edible starch therein, perhaps associated with more fiber. I have enjoyed the pithy middle portions of these in both autumn and spring. Early American eccentric Rafinesque (1828–1830) referred to 12 equally esculent species of arrowleaf, used for breadstuffs and soups, the roots useful for dropsy and yaws, the leaves applied to the breasts of nursing mothers to dispel milk. Cattle graze the leaves. Muskrats, por-cupines, and beavers gather the tubers, sometimes storing them in large caches. Ducks and geese eat the tubers and seeds, rails the seeds (Martin et al., 1951). CAUTION: Some marsh members of the jack-in-the-pulpit family also have arrowlike leaves, but their tubers and roots are dangerously loaded with oxalate crystals. One famous West Virginia herbalist once interjected at

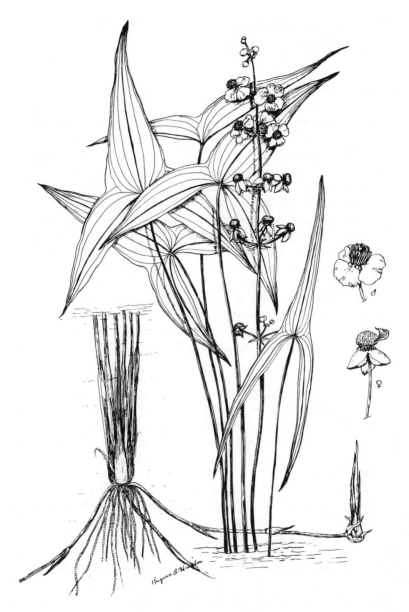

Sagittaria latifolia Willd.

one of my foraging lectures that the wapato nearly choked him to death. We discussed his mouth-boggling experience at length and finally concluded that he had probably gathered the poisonous arrow arum rather than the edible arrowleaf, arrowhead, or wapato. Be sure of your identifications! WSSA lists 9 species and/or varieties of *Sagittaria* as water weeds.

SAMBUCUS CANADENSIS L.   (CAPRIFOLIACEAE) --- American Elder, Elderberry

DESCRIPTION: Small shrubs arising from horizontal rootstocks, often forming dense colonies. LEAVES: two, opposite each other at the nodes, pinnate (i.e., divided into 5–11 leaflets, with one terminal and the others in pairs along the midrib or axis) or rarely decompound (with the lowermost leaflet itself subdivided); leaflets narrowly egg-shaped, 2–6 inches long, 1–2.5 inches wide, widest below the middle, with a tapering point at the tip, saw-toothed at the margin, basally rounded, with several lateral veins arising from the midrib like the plumes of a feather. FLOWERS: white to cream-colored, in large, terminal, flat-topped clusters 2–10 inches wide; sepals 5, small, basally united; petals 5, basally united; stamens 5; ovary 3-celled, with 3 terminal processes; united with the sepals. FRUIT: a small, purplish to blackish berry, usually 4-seeded, ca. 0.2 inch in diameter.

DISTRIBUTION: Common in meadows, fencerows, forest clearings, alluvial forests and swamp forests, growing but rarely bearing in the dense shade of the deciduous forests, flowering April through June, the first fruits not usually edible until late June in middle Maryland. SD-ME-TX-FL. Zones 4–10.

UTILITY: Fruits are available to eat out of hand from June, sometimes right on up to frost. Some taste good, some taste more medicinal. All are improved by conversion to jams, jellies, or wines, or added to other fruit mixtures, e.g., with sumac. Flower buds and unripe fruits are pickled like capers (Facciola, 1990). Drying the berries in the sun improves their flavor. The dried elderberry "raisins" can be eaten as is or added to stews, pies, and sauces. Like other foragers, I have dipped the flat-topped flower clusters in batter to make the so-called elderberry pancakes. More fastidious foragers strip off the flowers and add them to their batter, avoiding the more bitter flower stalks in the process. Flowers have also been used to make teas and wines. My long-standing friend Portia Meares (1981) gives several good elderberry recipes, one detailing how to make the wine. Made dry, it can substitute for burgundy, made sweeter for port. Her Elderberry Rob calls for 1 quart of elderberry juice, 1 teaspoon of lemon juice (or sumac, which see), and a teaspoon each of cinnamon, cloves, and nutmeg, plus later, 1 cup of sugar or honey. Hardcore foragers should resort to home-processed maple sugar. Maple sugar tappers sometimes removed the pith from elderberry stems and used them as spiles or spouts. Indians allegedly called elderberry the "tree of music", making their flutes thereof. Martin et al. (1951) record dozens of bird species which eat the fruits. Rabbits, squirrels, and woodchucks eat the fruits and bark. Chipmunks, mice, and rats eat the fruits. Deer, elk,

Sambucus canadensis L.

and moose graze the foliage, which may contain traces of cyanide. CAUTION: I am skeptical of recipes for pickled young flower buds and green fruits or for cooked young shoots. Young shoots and green fruits may contain cyanide. Some even recite tales of children poisoned by using peashooters fashioned from the hollow stems. Roots are also reported to be poisonous. My friend Dr. John Churchhill, MD (personal communication) can find no cyanide in American elderberries.

SCIRPUS ACUTUS Muhl. (CYPERACEAE) --- Hardstem Bulrush

DESCRIPTION: Tall wetlands weedy perennials, 3–9 feet tall when in flower, with rounded stems arising from horizontal, drab brown rootstocks. LEAVES: essentially lacking, or of green, blade-like flanges to 4 inches long, grasslike (very narrowly sword-shaped), with several equal parallel veins traversing the length of the "sword", pointed at the tip, toothless on the edges, basally flared out to clasp the stem. FLOWERS: hardly recognizable as such to the nonbotanist, in tight reddish to grayish brown clusters to the side just below the tip of the stem, individual florets embraced in green or brown scales; true sepals and petals absent; stamens usually 3; ovary 1-celled, free of the scales, with 2 or 3 terminal processes. FRUIT: a small, shiny, black, lens-shaped nutlet.

DISTRIBUTION: Aquatic or wetland species of fresh and slightly brackish waters, usually in full sun, beginning to flower in July, often flowering and fruiting until frost. WA-ME-CA-NC. Zones 3–7.

UTILITY: Most aquatic members of the sedge family CYPERACEAE might be sampled judiciously as foods, but how do you know the water in which they grow is not harboring amebas or giardia? The safest thing to do is to boil them vigorously, killing germs and a few nutrients as well. Indians used rootstock of this species, eating them raw, or pounding them into flour for breadstuffs. Montana Indians chewed the root to prevent or alleviate thirst (Moerman, 1986). I have eaten the rootstocks, but not made the bread. Yanovsky stated that even the pollen was used to make bread, like that of the rather similar cattail. The minute seeds likewise are said to have been used to make porridge or breadstuffs. Tanaka (1976) notes that young stems of *Scirpus fluviatilis* are peeled and eaten, like a stalk of sugarcane. Rhizomes of *S. lacustris* were eaten raw or made into breads by Amerindians, and the young shoots were consumed in spring. Even the rootstocks of *S. maritimus* were ground into an edible flour. Young budlings and roots of the Softstem Bulrush, *S. validus,* are eaten cooked or preserved in rice bran and the pollen is made into edible patties. Stem bases are eaten raw (Tanaka, 1976). Martin et al. (1951) rank the bulrush genus *Scirpus* as one of the most conspicuous plant groups in the American wetlands. The seeds constitute one of the more important and common foods of ducks, marsh and shorebirds. Stems and underground parts are eaten by muskrats and geese. As with grasses, sedges and bulrushes have not figured largely in the Amerindian pharmacopoeia. Moerman (1986) cites medicinal uses for only 7 of the many American species, none medicinally important. For abscesses, ceremony, consumption, crying children, hair growth, snakebite, sore throat, "spoiled saliva", and weak

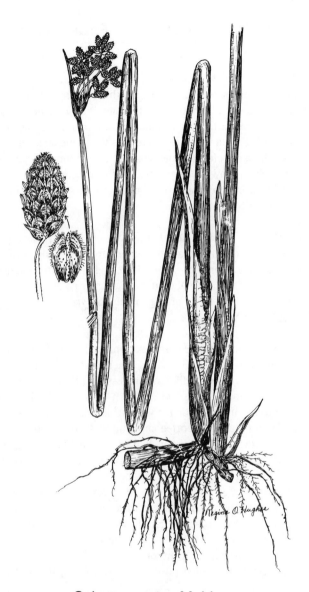

Scirpus acutus Muhl.

legs. That's about all the medicinal uses Duke (1986) and Moerman (1986) could dig up on *Scirpus*. WSSA lists 11 weeds in the genus *Scirpus*, all of which I would try for food in an emergency. CAUTION: Some of the species can cause ''grass itch'', and others can cut the unwary forager. A dermatitis attributed to *S. lacustris* may have been caused by an insect, the reed bug *Chilacis* (Mitchell and Rook, 1979).

SETARIA SPP.   (POACEAE) --- Foxtail Grass

DESCRIPTION: Erect or sprawling annual to perennial grasses, sometimes with horizontal rootstocks. LEAVES: grasslike, several at the base, the stem leaves one at a node, alternating on the stem, to 7 inches long, less than 1 inch wide, pointed at the tip, toothless at the margin, clasping the stem at the base. FLOWERS: in tight, elongate clusters at the apex of stalks that are usually longer than the clusters, the clusters with prominent hairs sticking out beyond the ovaries; flowers very much reduced, greenish, ripening golden brown. FRUIT: a minute grain, with many of these in the ripening flower clusters, all provided with long hairs.

DISTRIBUTION: Mostly annual weeds in old fields, waste places, some species growing in marshy situations, others growing in dry situations, flowering from May to July or later, some flowering and fruiting right up to frost. Ripe, fruiting heads can be gathered even after frost as they persist into the winter as attractive dried floral arrangements. *Setaria viridis* (Green Foxtail): WA-ME-CA-FL. Zones 3–8.

UTILITY: Like other grasses, for example, millets, these smaller grains can be eaten both in the "milk" stage when green or in the "grain" stage by the patient forager. One method of processing I have tried on the dry grass head is to burn it, thus getting rid of much of the chaff. Then the parched seed can be ground into a flour for use in making bread or porridge. Over-parched seed can be used as coffee substitutes. I grind mine in a mortar and pestle, but the Indian used rocks in lieu of mortar and pestle. One can also gather seeds by putting the tails of the foxtails on an old newspaper, the seeds tending to fall out with time. They can then be eaten out of hand, or prepared as a cereal, in all cases being careful to reject the stiff bristles sometimes attached to the seeds. Ground-up seeds make an interesting extender to corn-bread. One African tribe repelled rats from their stored grain by putting the sticky foxtail heads on top of the stores. Amaze your children with the magic foxtail. Put a ripe foxtail in your open shirt pocket upside down, having broken it off below the seeded portion. As you walk, the foxtail will climb out of your pocket!

> Foxtail, foxtail, where have you been?
> *Been down to China, and I come back again!*
> Foxtail, foxtail, where are you goin'?
> *Any old where that the wind is a'blowin'!*

Some *Setaria* species may yield 10 tons/acre biomass. As a weed in irrigated California alfalfa hay, foxtails may induce oral ulcers in cattle, reduce hay palatability, and decrease milk production. Folklorically regarded as antidotal

Setaria viridis (L.) Beauv.

(to varnish), antilactagogue, aphrodisiac, astringent, diuretic, emollient, laxative, refrigerant, and sedative, foxtail grasses have been used for bruises, cholecystosis, cholera, debility, dogbite, epistaxis, fever, nausea, pellagra, pregnancy, puerperium, pulmonosis, renosis, scratches, uterosis, vaginitis, etc. WSSA lists 11 species and/or varieties of *Setaria* as weeds.

SISYMBRIUM ALTISSIMUM L.   (BRASSICACEAE) --- Tumble Mustard

DESCRIPTION: Tall winter annual or biennial herbs, rarely to as much as 9 feet tall, often with a winter rosette. LEAVES: one at the node, alternating along the stem, the blades cut to the midrib, like a widely spaced, double-edged comb, the lobes narrow, untoothed, 1-nerved, progressively smaller up the stem, the rosette leaves to as much as a foot long and 3 inches wide. FLOWERS: yellow, in long clusters at the tips of the stems, the lowermost flowers opening first; sepals 4; petals 4; stamens 6; ovary 2-celled. FRUITS: of elongate, thin, rounded, long-stalked pods 1–4 inches long, with several seeds.

DISTRIBUTION: Weed of old fields, gardens, and waste places, flowering May to June, often dying back in sunny situations with the heat of summer. WA-ME-CA-FL. Zones 3–8.

UTILITY: Not finding this species in any of my handy foraging books (i.e., until Facciola, 1990), I bravely gathered it and ate it raw and as a potherb, indulging in all the aboveground parts of the plants, leaves, flowers, pod, stems, and roots, listed in decreasing order of edibility. Like other members of the family it is pungent, turning out about like turnip greens when boiled up as a potherb. Indians must have had time on their hands to gather and process the seeds, said to be a particularly good source of calcium, so much so that one vegetarian friend of mine is growing it as a calcium source. Though we planted it for him in the garden at Herbal Vineyard, that was before I knew what a common weed it was down on the government farm at Beltsville. We planted it too late. Like many other members of the cabbage family, it needs to be planted early to grow in the cool of the spring, or, conversely, it can be grown in the cool days of autumn. These cool-season annuals or biennials do poorly in the lowland tropics, unlike our hot-season annuals, which will appear as frequently in the tropics as they do here, maybe more so. Seeds can be sprouted or ground into gruels. Facciola (1990) adds that young leaves, stems, and flowers of *Sisymbrium irio* are eaten in salads. Young leaves and shoots of *S. officinale* are used in omelets, potherbs, salads, sauces, and soups. Ground seeds are used to make flour, porridge, or soups (Facciola, 1990). WSSA lists 4 more weedy *Sisymbrium, S. irio,* the <u>London Rocket</u> (I dislike that common name, as it suggests rockets like the equally edible and famous *Eruca* rocket); *S. loeselii,* the <u>Tall Hedge Mustard</u> (I dislike trinomials, even in common names); *S. officinale,* the <u>Hedge Mustard</u>; *S. orientale,* the <u>Oriental Mustard</u> (I dislike that common name, as it sounds too much like a *Brassica*); and *S. sophia,* which is better known now as *Descurainia sophia,* according to WSSA. But I should not complain about a

Sisymbrium altissimum L.

problem, unless I can offer alternative solutions to the problem. Having worked for a while with the WSSA nomenclature people, back in the days of George Darrow, I would recommend the compounded word <u>Hedgemustard</u> as the generic common name, preceded by London, tall, official, and Oriental as qualifiers.

SMILAX ROTUNDIFOLIA L.   (LILIACEAE) --- Catbriar, Greenbriar, Horsebriar, Roundleaf Catbriar

DESCRIPTION: High-climbing, green-stemmed, thorny perennial vines, the rootstocks slender. LEAVES: one at the node, alternating along the stem, deciduous or sometimes evergreen farther south, narrowly to broadly egg-shaped or almost round, 2–5 inches long, 1–4 inches broad, pointed at the tip, toothless or occasionally spiny-toothed on the margins, the bases tapered, rounded or heart shaped at the attachment of the leafstalks, with 3–7 veins arising there, arching out and back in towards the tip. FLOWERS: greenish or bronze-colored, in few-flowered, flat-topped clusters in the angle formed by the leafstalk and stem; sepals 3; petals 3; stamens 6; ovary 3-celled, with 3 terminal processes. FRUIT: a bluish black, globose drupe or berry 0.2–0.4 inch in diameter, with 1–3 seeds.

DISTRIBUTION: Common weeds in sunny and shady thickets and moist deciduous forests, fencerows, and waste places, flowering from April through June, the fruits ripening from September to frost. NY-ME-LA-FL. Zones 5–9.

UTILITY: I eat the young, tender shoots of any greenbriar I happen to encounter on a hungry hike. This is our most common species and one can find fresh shoots at the tips of branches far into summer. The teas I have made from the knobby roots have not been noteworthy and give no evidence of containing testosterone. Peterson notes that the young shoots, leaves, and tendrils are good prepared like spinach or added to salads. Bullbriar (with arrow-shaped leaves) roots are the source of a gelatinous material used to thicken stews and root beers, etc. We might add any of the temperate-zone *Smilax* roots to our root beers for a poor man's sarsaparilla. Tropical species are more appropriately the source of sarsaparilla, claimed by some to contain the hormone testosterone. You might have seen advertisements in subpor-nographic magazines with the hint that sarsaparilla makes a man much more virile. I am pleased to have transplanted from the forest to Herbal Vineyard a pair of plants which I call ''macho/hembra'' (= male/female). I found a species of *Smilax* near home, twining around *Dioscorea villosa,* the wild yam, and mother of the contraceptive pill. Both plants, though not closely related, do contain steroid hormones, the *Dioscorea* surely containing dios-genin, the *Smilax* probably not containing testosterone. Both probably contain traces of boron in their roots, especially if grown out west where boron is more prevalent. According to one USDA study, 3 milligrams, about 1/10,000th of an ounce, of boron can double blood levels of estrogen and testosterone.

Smilax sp.

Cherokee and Koasati Indians scratched themselves with this briar for local cramps, pains, like headaches, and twitching (Moerman, 1986). WSSA lists 7 more weedy *Smilax, S. auriculata,* the Wild Bamboovine; *S. bona-nox,* the Catbriar, *S. glauca,* the Cat Sawbriar; *S. laurifolia,* the Bamboovine; *S. pumila,* the Sarsaparillavine; *S. smallii,* the Lanceleaf Greenbriar; and *S. walteri,* the Coral Greenbriar. I would not be afraid to cautiously sample the growing shoots of any of these, before the spines have hardened on the spiny species.

SONCHUS OLERACEUS L. (ASTERACEAE) --- Annual Sowthistle, Common Sow Thistle

DESCRIPTION: Hollow-stemmed annual herbs to 6 feet tall, with a cream-colored juice exuding from breaks in the stems. LEAVES: one at the node, alternating on the stem, dandelionlike, 2–12 inches long, 0.5–6 inches wide, rounded or sharp-pointed at the tip, deeply lobed, the lobes themselves sometimes saw-toothed, the teeth sometimes spinelike, the leaves clasping the stem at their base, with no obvious leafstalk. FLOWERS: yellow, small, aggregated into dandelionlike heads, these arranged in round-topped terminal clusters, the yellow flowers being replaced by silky fuzz (actually part of the fruits) with age. Fruit a small 6- to 10-ribbed nutlet, 0.1–0.2 inch long.

DISTRIBUTION: Very common weed in waste places and cultivated fields, pastures, and roadsides, flowering from June until frost in Maryland, often much earlier farther south. WA-ME-CA-FL. Zones 3–8.

UTILITY: The other annual species, *Sonchus asper,* has rounded rather than angular clasping bases on the lower leaves. The perennial species, *S. arvensis,* has horizontal rootstocks. All are equally edible, or inedible, depending on your point of view. Even after two changes of water, the young shoots are bitter and the older shoots are more bitter and tougher to boot. If I have to eliminate a species from this book, this is a good candidate for scrubbing. Still, western Indians used *S. asper* as a potherb, and Romans used it both as salad and potherb. Europeans still add the tenderer leaves to soups. According to Tanaka (1976), Javans eat the whole plant of *S. arvense,* putting the leaves in curries or rice dishes. Young tops and leaves, even of the spiny *S. asper,* are steamed and eaten in Asia. One 13th century herbalist probably stimulated wider use of the herb when he recommended a diet of sowthistle "to prolong the virility of gentlemen" (Mitich, WT2:380.1988). (The doctrine of signatures may have led many plants with milky sap to be considered to prolong the virility of gentlemen.) I find the flowers of any of the Maryland species quite tasty when chewed, quickly turning into a gummy cud, pleasant to the taste. (See the table under *Rosa.*) I suspect they might be more effective against jaundice and cirrhosis than against impotency. The doctrine of signatures also suggested that ingestion of the herb would stimulate the flow of milk in lactating mothers. Cosmetically, it was suggested that sowthistle also cleared a woman's complexion. British people value the herb for veterinary purposes. Animals often relish the weed, having lost interest in the other forages available (WT2:380.1988). Maoris of New Zealand make a chewing gum from the white milk. Roots of some are used as vegetables

Sonchus oleraceus L.

or scorched and used as coffee substitutes. WSSA lists 3 more weedy species of *Sonchus: S. arvensis,* the <u>Perennial Sowthistle</u>; *S. arvensis* ssp. *uliginosus,* the <u>Marsh Sowthistle</u>; and *S. asper,* the <u>Spiny Sowthistle</u>.

SORGHUM HALEPENSE (L.) Pers.   (POACEAE) ---
Johnsongrass

DESCRIPTION: Tall, erect, tufted or clumped, bamboolike perennial grasses, from knotty horizontal rootstocks, from which grow many fibrous roots; to 3, rarely 6 feet tall; stems green, hairless. LEAVES: basal and then one per node, alternately or spirally arranged on the stem, the blades 8–20 inches long and to 1/2 inch broad, rarely hairy at the junction of leaf and stem, or rough to the touch (scabrous) at the edges; with no obvious leafstalk and no teeth; clasping where the leaf joins the stem; veins parallel. FLOWERS: numerous, tawny or purplish, loosely scattered in plumose terminal clusters up to 20 inches long, 6 inches broad. FRUITS: small, contained within the minute, leaflike floral parts, the grain reddish brown, only ca. 1/10 inch long.

DISTRIBUTION: Sometimes forming solid stands in meadows, pastures, roadsides, swamps, and waste places. Flowering May in Carolina to July in Maryland, often fruiting until frost. WA-MA-CA-FL. Zones 5–9.

UTILITY: Only young shoots are palatable as forage for cattle and foragers, and might yield forth a hard-earned starch or sugar. Young rootstocks and stalks might serve as nibbles before flowering. Rootstocks may also be sun-dried, ground, and beaten into flour. Rootstocks might provide food all year round, roasted or boiled. The grains can be used as a poor man's cereal. Larry Mitich (WT1:112.1987) details the lamentable introduction of this weed. Like all sorghums, he says, it occasionally causes cyanide poisoning in cattle. It is named after Col. William Johnson from Selma, AL, who shortly after 1840 sowed it on his Alabama River valley farm. As early as 1849, an Arkansas farmer bragged of his African johnsongrass imported from Jamaica, giving 2 tons of hay per acre, four cuttings a year. If cut before seeding, it can yield nicely. Left alone, it can readily get out of hand. Swine, very fond of the roots, might be enlisted to help eradicate infested areas. Other species of *Sorghum* are widely cultivated for sugar, even for brooms. Thinking back 55 years to rural Alabama, I remember riding in an ox-cart with my dad and one of his countrier cousins, and they were talking about sorghum molasses. You see, in temperate Alabama, you could grow a crop of sorghum, but not of sugarcane. I believe my earlier pancakes were topped with homemade sorghum syrup. Even today, a few places in the southern U.S. grow sorghum for brooms as well. The Shattercane, a weedy variant of sorghum, *S. bicolor,* might well be used just like sorghum, as a source of cereal and sugar. CAUTION: Some species of *Sorghum* are notorious for cyanide production. Watt and Breyer-Brandwijk (1962) relate how seed of the South African sorghum, known as kaffir corn, is edible and is used to brew kaffir beer. But the plant is toxic to cattle, especially when wilted, or when cut after drought or frost.

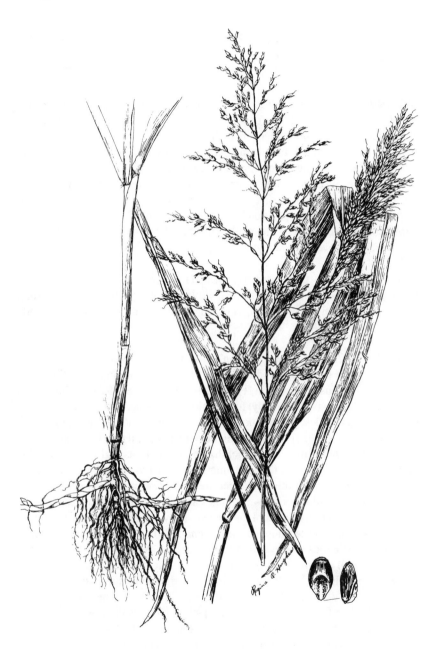

Sorghum halepense (L.) Pers.

It seems that stress of many types might induce cyanide production, running to as high as 50–3000 ppm. Poisonings are known from *Sorghum* on almost every continent, e.g., in India, Egypt, and North America.

STELLARIA MEDIA (L.) Cyrillo   (CARYOPHYLLACEAE) --- Common Chickweed

DESCRIPTION: Clambering, weak-stemmed annual, rarely perennial herbs, the stems up to 2 feet long. LEAVES: 2 at the node, opposite each other, egg-shaped, rarely more than an inch long, narrower, pointed at the tip, toothless on the margins, basally tapered or rounded or notched to the leafstalk, this sometimes lacking; midrib with a few weak veins arching out towards the edges. FLOWERS: white, often twinned in few-flowered clusters, at the tips of the branches, or arising in the angle of the leaf with the stem; sepals 5; petals 5, the tips deeply notched; stamens 2–10; ovary 3-celled, with 3 terminal processes. FRUIT: a small, egg-shaped pod splitting from the tip to the base, with several seeds.

DISTRIBUTION: Common early spring or autumn weed of gardens, lawns, waste places, occasional in the forests, flowering and fruiting in early spring, and again (probably new seedlings) in fall. Available through much of a mild winter. WA-ME-CA-FL. Zones 3–9.

UTILITY: If so many people didn't assert that this would help you lose weight, I would have excluded chickweed from this foraging book. To avoid being myself accused of sexism, I quote Mrs. M. Grieve (1974): "Chickweed water is an old wives' remedy for obesity." Any herb that can help slim obese Americans deserves a spot. So far, however, it has not helped Jim Duke. I suspect that if Jim ate nothing but chickweed, it would certainly have a slimming, if not downright demoralizing effect. Elias and Dykeman think higher of chickweed than I, suggesting that mild-flavored chickweed is excellent by itself or with other stronger greens. They suggest its generous use raw in salads or boiling only 2–5 minutes and adding it to other, less delicate greens near the end of their cooking period. They suggest serving with butter, seasonings, and a little chopped onion, enough of which will make almost any potherb delicious. Steve Brill recommends the stems as well as the leaves. Facciola (1990) mentions that Indians used the tiny seed in breads and soups, a use not mentioned by Yanovsky. Facciola mentions that natural-food enthusiasts liquefy the chickweed, sometimes clipping off the tops of young plants, allowing them to regrow. Jones (1991) mentions the use of chickweed for 9:00 a.m. in floral clocks. But she warns us: "chickweed can produce five generations of chickweedlets in a single season". Dozens of species of birds feed on the seed, blackbirds, buntings, doves, finches, juncoes, larks, pipits, siskins, sparrows, and towhees feed on the seed, while rabbits and sheep graze the weed (Martin et al., 1951). As its name indicates, it was once valued for chickenfeed. According to Grieve (1974), pigs like it, cows and horses will eat it, sheep are indifferent to it, while goats refuse to touch

Stellaria media (L.) Cyrillo

it. CAUTION: Kingsbury lists this as a plant accumulating levels of nitrates toxic to grazing cattle. WSSA lists 4 other weeds in the genus *Stellaria*, *S. aquatica*, the Water Starwort (under the alternative name *Myosoton aquaticum*); *S. graminea*, the Little Starwort; *S. holostea*, the Easterbell Starwort; *S. laeta*, the Alaska starwort; and *S. pubera*, the Star Chickweed. Only *S. media* is important to wildlife.

TARAXACUM OFFICINALE Wiggers (ASTERACEAE)
--- Dandelion

DESCRIPTION: Stemless (except for the "flower stalks") biennial or perennial herbs, with white milk exuding where broken, even in the deep taproot. LEAVES: all in the basal rosette, so characteristic that I have elsewhere used dandelionlike as a descriptive adjective meaning "long with jagged, irregular lobes, often broadest above the middle, much narrower or reduced to the midrib closer to the base, slightly pointed at the tip and the tips of the lobes". FLOWERS: very small, aggregated into yellow heads that the amateur takes to be the flower, the heads on long stalks 1–12 inches long; sepals 5, basally united; petals 5; stamens 5; ovary 2-celled, with 2 terminal processes, united with the sepals. FRUITS: of slender brownish nutlets, each with a "parachute" at the tip, with dozens of these in each flower head after it goes to seed.

DISTRIBUTION: Perhaps our most famous and persecuted, if not most common, lawn weed, the dandelion is an opportunist that is liable to pop up in any opening anywhere other plants can carve out a niche, flowering as early as February down south to June farther north, and sporadically flowering right up to frost (sometimes all winter in Maryland). WA-ME-CA-FL. Zones 3–9.

UTILITY: I have eaten, not necessarily enjoying, all parts of the dandelion; the large taproots, boiled and pickled, are quite nice, and can be found most of the winter in protected situations; the greens are equally available, but they are bitter to my taste unless vigorously doctored up; I have also deprived goldfinches of the chewy seeds which afford some sustenance and no doubt B vitamins and sterols. For one 2-week period, I consumed at least 100 flowers every day for lunch, to see if the high carotenoid content would turn me yellow. Results negative! The yellow flowers, perhaps reflecting the doctrine of signatures, have been used folklorically for jaundice and other liver ailments. Recently it was reported that soy lecithin prevented cirrhosis in chimpanzees. Dandelion flowers were reportedly higher in lecithin than soy. Saner people than I make delectable wines from the flowers, adding sugar, etc. Some people cook up pancakes with unopened flower buds in the batter. One Arabic cake called *yublo* also contains the flowers (Facciola, 1990). I suspect the seeds could be bagged up with their parachutes, burned, parching them in the process, and ground up into flour. I'll save that experiment for next spring unless I have a good autumn haul of dandelions. Facciola mentions that the sprouts are edible. Anglo-Saxons, Celts, Gauls, and Romans all indulged in the dandelion as food. Legend has it that Theseus indulged in the dandelion salad following his slaying of the Minotaur (WT3:537.1989).

Taraxacum officinale Wiggers

Indians used the leaves as greens, cooked with water, vinegar, or with meat, and apparently used the roots as salads. Perhaps the latest champion of the dandelion is Peter Gail, who has an interesting series, *ON THE TRAIL OF THE VOLUNTEER VEGETABLE,* in the trade journal, *THE BU$INESS OF HERBS,* wherein he has written engaging articles on such weeds as dandelion, daylily, lambsquarter, and purslane. Jones (1991) notes that the U.S. imports more than 100,000 pounds of dandelion annually for use in patent medicines.

THLASPI ARVENSE L. (BRASSICACEAE) ---
Fanweed, Field Pennycress, Frenchweed, Stinkweed

DESCRIPTION: Erect, hairless, annual herbs, rarely branching, to 2 feet tall, with a basal rosette of ephemeral leaves. LEAVES: of the rosette willowlike, broadest above the middle, with a short stalk; stem leaves one at the node, alternating on the stem, smaller, stalkless, to 4 inches long, toothless or toothed, the leaf base clasping the stem arrowheadlike; midrib with a few veins arching out towards the margins like plumes of a feather. FLOWERS: in elongate clusters at the tip of the stem, white, the lowermost flowers opening first, such that you have ripe fruits at the bottom, green fruits in the middle, and flowers at the top; sepals 4; petals 4; stamens 6; ovary 2-celled, with a single terminal process. FRUIT: a flat, green or brown pod notched at the tip, 0.3–0.6 inch wide, with about 12 blackish seeds.

DISTRIBUTION: Barnyards, old fields, and waste places, often a troublesome weed, flowering as early as March in the Carolinas, in April in Maryland, often dying back with the heat after fruiting. I suspect that there is a spring crop and a separate smaller fall crop of this species. WA-ME-CA-FL. Zones 4–8.

UTILITY: I have eaten all the pungent parts of this plant in spring, flowers, leaves, pods, stems, and roots, in decreasing order of palatability. All are improved by boiling in two changes of water, at that point approaching turnip greens in flavor and pungency. Ground seed pods could be used as a pepper or mustard substitute by those with more vivid imaginations than mine. And, as with many members of this family, the seeds can be sprouted. Still, I have used all the parts, just like the ancient physician, Mithridate, who tried his concoctions on himself. CAUTION: Having earlier mentioned the virtues of the allyl isothiocyanates as cancer preventives, let me advise you what overdoses can do to cattle who ingest too much, clearly resulting in gastric distress. Further, milk from such cows may have a garlic flavor. Kingsbury says that, undoubtedly, allyl isothiocyanate, our cancer preventive, is responsible for reports of gastric distress in livestock feeding upon grain contaminated by the "fanweed seed". Fanweed was also shown to be responsible for blood disorders in Washington State heifers who had grazed hay contaminated with the fanweed. Their feces still revealed passing fanweed seed. Trouble ceased when noncontaminated hay was substituted for the contaminated hay. Cattle ingesting too much seed may suffer abortion, diarrhea, enteritis, hematuria, nephritis, and cardiac and respiratory paralysis. Having cited so much negative data from a poisonous plant book and a weed manual, how can I go on suggesting these things as food? I really want to discourage novice grazers from overgrazing. I will continue to sample these seeds in moderation. After

Thlaspi arvense L.

several soul-searching years, I have learned nothing to suggest that small doses of poisons are or are not okay, maybe even healthy. The Thoroughwort Pennycress, *Thlaspi perfoliatum* L., can serve for food, in moderation as well.

TRAGOPOGON DUBIUS Scopoli   (ASTERACEAE) ---
Western Salsify, Yellow Goatsbeard

DESCRIPTION: Biennial or perennial herbs to 3 feet tall, with a thick taproot, a milky liquid coming out at wounds in the plant. LEAVES: one at a node, alternating up the stem, grasslike, 4–12 inches long, 0.2–0.5 inch wide, pointed at the tip, toothless on the margins, clasping the stems at the base, with parallel veins extending the length of the leaf. FLOWERS: small, yellow, in dandelionlike heads at the ends of long terminal stalks, the stalk hollow and swollen just below the flower head. FRUITS: of long, round, ribbed nutlets with a terminal, dandelionlike parachute nearly an inch long.

DISTRIBUTION: Roadsides, sandy waste places, railroad rights-of-way, flowering from April (North Carolina) to June (Maryland), the fruits ripening quickly like a dandelion. Flowers are open only briefly in the morning, usually closing by 11 o'clock or noon. WA-NY-CA-NC. Zones 3–7.

UTILITY: Some people prefer the roots of the goatsbeard to the larger, cultivated salsify or oyster plant or vegetable oyster, *Tragopogon porrifolius,* of the same genus. Facciola (1990) says of the western salsify that the flowering stems, including the flower buds, are cooked and eaten like asparagus. Young leaves, shoots, and diced roots can be used in salads. Fully developed roots of the wild plant may be blanched, peeled, cooked, and eaten like the salsify itself (Facciola, 1990). Conversely, Watt and Breyer-Brandwijk (1962) maintain that the wild one is inedible. Roots of common salsify are used raw in salads, or more often boiled, baked, or fried in butter, even ground up and added to cakes. Facciola (1990) adds that young shoots (called chards), flowerbuds, and flowers, for the flower lovers, may be eaten in salads, or they too may be cooked and eaten. According to Peterson, "young stems when a few inches high and the bases of the lower leaves make a delicious cooked vegetable." Peterson suggests parching the roots of the salsify, much as we do with the related chicory, as a substitute or adulterant for coffee. I tried eating the seeds raw as I have those of the dandelion, but the tough parachute made that an unpleasant experience. Next summer, I will burn the parachutes, burning off the chaff while parching the cereal. Then I will grind the parched seeds in my mortar and pestle. I expect a high-fiber meal that will make a crunchy bread. Perhaps it would be more rewarding to sprout them for use in salads and sandwiches, as suggested by Facciola (1990). British Columbian Indians used the coagulated milky juice of some tragopogons as chewing gum. Like dandelion and many other white-sapped members of this family, salsify contains rubber, ca. 1000 parts per million. Australians claim that cattle will

Tragopogon dubius Scopoli

graze the plant, but not when it is in flower (Watt and Breyer-Brandwijk, 1962). Navajo Indians used the salsify as a ceremonial emetic and in the treatment of dogbite or coyote bite (Moerman, 1986), the meadow salsify for boils, throat ailments, and for internal injuries to their horses. WSSA lists 3 weed species in *Tragopogon* which probably can be used interchangeably: *T. dubius*, the Western Salsify; *T. porrifolius*, the Common Salsify; and *T. pratensis*, the Western Salsify.

TRIFOLIUM PRATENSE L. (FABACEAE) --- Red Clover

DESCRIPTION: Biennial or perennial herb to as much as 2 feet tall, the stems branching, locally hairy. LEAVES: one at the node, alternating on the stems, "cloverlike" (composed of three similar leafets), the leaflets egg-shaped, to 2 inches long, 1 inch broad, usually broadest below the middle, rounded or slightly notched at the tip, finely saw-toothed on the margin, tapered to the base, with numerous lateral veins arching out towards the margins, the leafstalk of the compound leaf often much longer than the leaf. FLOWERS: pink, pealike, in short-stalked, tight clusters close to the upper leaves; sepals 5, united basally; petals 5; stamens 10; ovary 1-celled. FRUIT: a small, 1-seeded pod, the top coming off like a hat when ripe.

DISTRIBUTION: Planted in pastures and volunteering in fields and pastures, sometimes very common along superhighways, flowering as early as April in the Carolinas, May in Maryland, the red clover's fruits are of little interest to foragers, but continue to ripen up to frost. WA-MA-CA-FL. Zones 3–8.

UTILITY: The best forager's bouillon I ever created consisted of flowers of red clover, chicory, and wild onion, with salt and pepper. Harris (1971) substitutes the leaves for lettuce leaves in his cheese sandwiches. Moreover, he recommends a clover blossom tea with mint and dandelion as a good summer substitute for the cold-weather chickweed tea for those on reducing diets. Orientals are said to eat the leaves. Powdered leaves and flower heads are sprinkled onto rice (Facciola, 1990). Japanese cook them in soy sauce or put them in soups (Tanaka, 1976). The tiny seeds can be winnowed and ground up for a famine flour. (Some people prefer clover sprouts to alfalfa sprouts.) Peterson suggests grinding up the whole flower heads as a breadstuff, a much simpler and probably better balanced foodstuff. He also suggests the dried flower heads for teas. The flower heads, with a strong folk reputation for helping cancer, apparently contain estrogens. In *AN HERB A DAY...RED CLOVER* (Duke, 1990), I summarized recent research suggesting that the isoflavones may block breast cancer at an early age (age 15–25). The compound *genistein* "might give the immune system a better shot at destroying" cancer cells (*Science News*, May 12, 1990). It's good to read that in *Science News*. Ten years ago, the Quack Busters suggested that anyone talking about regulating the immune system was a quack. Fifteen years ago, I listed clover in my tongue-in-cheek Quack Salad (Duke, 1977). Today we hear top scientists talking about isoflavones like genistein regulating the immune system and possibly preventing the spread of breast cancer. When doctors hinted I might have intestinal cancer (in the fall of 1986), I started drinking a lot of

Trifolium pratense L.

clover-flower tea, made more acid with *Oxalis* or *Rumex* leaves or both. Only in 1990 at an NCI soybean symposium did I learn that the legume isoflavones were both estrogenic and apparently cancer preventive, vindicating the long-term reputation of red clover as a folk cancer remedy. This is one of the more important clovers for our eastern wildlife, feeding many birds and mammals. WSSA lists 9 weedy species of *Trifolium*.

TUSSILAGO FARFARA L. (ASTERACEAE) ---
Coltsfoot, Coughwort, Horsehoof

DESCRIPTION: Strange, stemless perennial, the flowers appearing alone in spring, followed by the leaves, all emerging from an underground rootstock and all usually less than 1 foot tall. LEAVES: emerging from below ground or from a clambering, telescoped, gray-green stem, like rounded, heart-shaped, gray-green umbrellas, the blades 2–12 inches long, 2–10 inches wide, barely pointed at the tip, notched at the leafstalk, scalloped in between, white underneath due to a mat of wooly hairs. FLOWERS: yellow, small, aggregated into solitary, dandelionlike heads at the tip of grayish or whitish ghostlike stalks. FRUITS: yellow flowers are replaced by silky fuzz (actually part of the fruits) at the tip of miniature seeds much like those of the dandelion, a small, 6- to 10-ribbed nutlet, ca. 0.1 inch long.

DISTRIBUTION: Common weed on landslides, mine fills, roadsides, and shale barrens. Flowers appearing before the leaves, as early as March in Maryland, the fruits parachuting away in April or May. MN-ME-WV-MD. Zones 3–6.

UTILITY: Impure poetic license induced me to include coltsfoot, an herb villified with the FDA classification of ''Herb of Undefined Safety''. Perhaps undefined safety is better than no safety at all. I think of coltsfoot as more medicinal than food. Still, I include it as a reminder of that wonderful forager's newsletter by the same title, *COLTSFOOT*. Put out in Virginia 6 times a year, *COLTSFOOT* deals with many of the same issues covered in this book. As a matter of fact, an item followed by a parenthetical (cf) can be found in *COLTSFOOT*, which was my source for that bit of information. I have made coltsfoot candy, boiling down the rootstock and adding sugar. Pleasant to the taste, coltsfoot candy may have helped my cold, as expectorants often do. The ancient Greek name for the plant meant ''cough plant'' (Jones, 1991). I have not smoked the leaves for asthma, not eaten the leaves as a vegetable as Europeans were wont to do. I have drunk coltsfoot tea with pleasure, only after lemon and sugar were added to taste. With its strange, ghostlike flowers, developing long before the leaves emerge, coltsfoot can be important in both your floral clock and floral calendar gardens. In January of 1991, my first coltsfoot flowered January 15, and there were still flowers, but no leaves, March 15. The early coltsfoot beat my crocus by a month. Writing this book has kept me away from home most days. But I know that flowers of coltsfoot, like daylily, dove's dung (*Ornithogalum*, excluded herein because I think it unwholesome), evening primrose, and salsify are good for floral clocks be-

Tussilago farfara L.

cause they are open only certain hours of the day (or night with the evening primrose). Jones (1991) describes coltsfoot as ''a confirmed sun worshiper. Its flowers open only when the sun shines fully on them; they stubbornly refuse to unfurl on overcast days, just as they routinely close up at dusk.'' CAUTION: Like comfrey (*Symphytum*) and squawweed (*Senecio*), coltsfoot contains pyrrolizidine alkaloids. Rats with diets containing more than 4% coltsfoot developed cancerous afflictions of the liver. (Duke, 1985)

TYPHA SPP.   (TYPHACEAE) --- Cattail

DESCRIPTION: Perennial aquatic or marsh perennials, to as much as 15 feet tall from a horizontal rootstock. LEAVES: to 6 feet long, one at the node, alternating on and embracing the stem, stalkless, grasslike, pointed at the tip, toothless on the margin, with several parallel veins running from the base nearly to the tip. FLOWERS: hardly recognizable as such, more recognizable as the clublike cattail, probably as familiar to herbal novices as acorns and pine cones; male flowers in green thickenings at the tip of the central stem, shedding pollen copiously for a while, some of it reaching the thicker female swelling below it on the same stem. FRUITS: tiny, hairy nutlets, by the hundreds in the dark brown female portion of the cattail.

DISTRIBUTION: Common in roadside ditches, swamps, ponds, and sluggish, sometimes brackish, streams, flowering from May to July, when most of the pollen is long gone, the fruits ripening until and past frost. WA-ME-CA-FL. Zones 3–10.

UTILITY: One of our easiest foraging species to recognize and one of the most provident, recognizable and providing food all year long. One June 7, I shook the male parts of the spikes into a paper bag, easily obtaining an ounce of pollen (and half an ounce of thrips) from 25 shakedowns. This pollen is mixed half and half with other flours to make johnnycakes. At this time of year both male and female parts of the spike are boiled and eaten like corn on the cob. I've eaten the female parts raw, finding them chewy but not unpleasant. In fall and spring, one can find the edible spears arching out of the mud; the inner pith of the horizontal rootstocks has been tasty every time I have tried it. Indians made flour from these roots, said to attain incredible yields of 140 tons/acre (Harris, 1971), yielding ca. 32 tons of flour. Even the minute seeds were eaten by the Indians, who apparently burned the fuzz, thus eliminating the chaff and parching the seeds, which then could be added to porridges or breadstuffs. Pioneers used the stout stems to hold wicks in candlemaking. Seeds are so small as not to be overly attrative to even birds. On occasion, sandpipers and teal have engorged on the seeds, geese on the underground stems. As important food sources, cattails are one of the best plants for muskrat marshes, the muskrat, like foragers, very fond of the starchy underground stems. Since cattails have little value for ducks, cattails need managing on duck-hunting preserves. Mowing the cattails after the "tails" have formed, but before they have browned and turned ripe, followed by a second mowing about a month later, when growth is 2–3 feet tall, will "kill

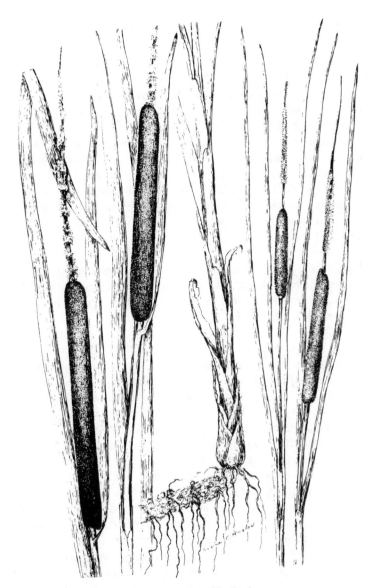

Typha angustifolia L.

at least 75% of the plants'' (Martin et al., 1951). WSSA lists *Typha angustifolia* (Narrowleaf Cattail), *T. domingensis* (Southern Cattail), *T. glauca* (Blue Cattail), and *T. latifolia* (Common Cattail) as weeds, valuable though they be.

URTICA DIOICA L.   (URTICACEAE) --- Stinging Nettle

DESCRIPTION: Perennial herb to nearly 6 feet tall. LEAVES: 2 at the node, opposite each other on the stem, the lower leaves 2–5 inches long, 1–3 inches wide, broadly egg-shaped, broadest below the middle, with a point drawn out at the tip, coarsely toothed along the margin, with a heartlike notch where the leafstalk attaches to the blade, with 3–5 nerves arising where the leafstalk joins the blade, the upper leaves smaller and proportionately narrower. FLOWERS: greenish, in clusters in the angle formed by the leafstalk and the stem, longer than the leafstalks, the male and female on different plants; male flowers with 4 sepals and 4 stamens; female flowers with 4 sepals and a 1-celled ovary free of the sepals. FRUIT: a small, 1-seeded nutlet embraced by 2 expanded sepals.

DISTRIBUTION: Rich, alluvial meadows and forests, often forming dense stands, spreading by underground rootstocks, flowering from May to July in the Carolinas, from June to July in Maryland, the fruits ripening from July to September. WA-ME-NM-FL. Zones 3–8.

UTILITY: I am probably the only forager you'll read about who has eaten the stinging nettles raw, although most foragers have enjoyed them cooked. (Julie Summers may have reported a raw nettle repast somewhere in the pages of *COLTSFOOT.*) The raw nettles quit stinging by the time they get to the throat, at least in my trials. The nettles are easy enough to secure in summer as vitamin-rich potherbs. Once you know where the solid stands are, you can try an old Scotch trick. Go dig up the roots when the ground is thawed in winter, taking them to a dark cellar. It won't be long before you have blanched nettle to mix with your blanched pokeweed. I hardly need to do that in Maryland, at least under global warming. My nettles persisted in to January 1991, on the south side of the barn at Herbal Vineyard, and new shoots were up the following months, in a warm winter of the warmest year on record. Nettle juice can be used to coagulate milk. I have read that different ethnic groups, scattered all over the world, use the stinging hairs as a counterirritant in arthritis. Several people I know who tried this now keep nettles in their kitchen window beside the aloe. The mother of my fiddle player no longer needs the nettle, though it is now a weed in her garden. Her arthritis went away. My degenerative spinal arthritis does not seem to respond. Heinerman (1988) suggests a slenderizing Nettle Ale, boiling down 2 gallons water, 4 quarts fresh nettle greens, 2 lemons, 3 limes, 2 ounces ginger root, 2 cups brown sugar, a little nutmeg and mace, primed with a cake of active yeast. Euell Gibbons recommended nettle for obesity (Heinerman, 1988). Most medical doctors would advise against self-diagnosis and medication, let alone self-urtication. But I recently learned that the sting of the nettle simulates the

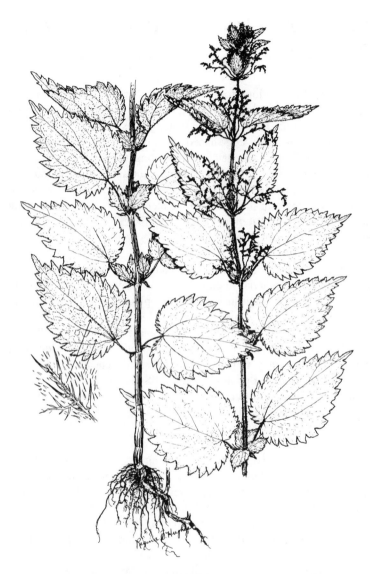

Urtica dioica L.

effect of acupuncture, electrostimulation, or ultrasound at alleviating the cramps and pain associated with a herniated disk. Any of these three will cause twitching in the cramped muscles and numb limbs that have accompanied my slipped-disk problem. WSSA lists 3 other weeds in *Urtica, U. chamae-dryoides,* the Southern Nettles; *U. lyallii,* the Lyall Nettle; and *U. urens,* the Burning Nettle.

UVULARIA PERFOLIATA L.   (LILIACEAE) --- Mealy Bellwort

DESCRIPTION: Colony-forming perennial herbs to 24 inches tall, the stem unbranched or with one branch, from a strikingly white, horizontal rootstock. LEAVES: one at the node, alternating along the stem, 1–5 inches long, 0.5– 2 inches wide, narrowly egg-shaped, pointed at the apex, toothless on the margin, the leaf base completely surrounding the stem (hence perfoliate), with 3–5 veins arising at the base, arching out and back in towards the tip. FLOWERS: yellowish green, drooping, stalked, about 1–1.5 inches long, solitary in the angle of the leaf with the stem; sepals 3; petals 3; stamens 6; ovary 3-celled, with 3 terminal processes. FRUIT: a 3-angled, few-seeded pod, broader at the tip than the base.

DISTRIBUTION: Moist, deciduous woodlands, almost always in the shade, often in coves and alluvial floodplain forests, flowering from April to June, fruiting from June to August. I suspect this plant is more ephemeral than the books would indicate. I have some tagged specimens to watch. I lived at Herbal Vineyard and walked by *Uvularia* on the way to my ginseng patch at least 10 years, but never noticed the *Uvularia* until 1986. I would have missed the fruits because they were so well concealed beneath the leaves. But I collected a specimen for photographer Neil Soderstrom, and spotted the fruit (of related *U. sessilifolia*) only as I was delivering the trophy, with splendid white rootstock, to Soderstrom. OH-MA-LA-FL. Zones 5–8.

UTILITY: Tanaka (1976) says that this species is used in a diet drink, quoting Fernald and Kinsey (1958) quoting one Mannaseh Cutler back in 1785. I didn't know they needed diet drinks way back then. Uphof (1968) speaks of the young shoots as an asparagus substitute, and the roots as edible when cooked. Until 1976, I never sampled the bellwort. That year I enjoyed the root as a tender nibble, but sent most of it off for analysis. Now that I know where it is, I'll be lurking for those ''asparaguslike'' spears when spring comes next year. Peterson recommends discarding the leaves off the spears as they turn bitter in the cooking, like those of solomon's seal. As the plants are small and not common, they should be used only in case of emergency; I'll propagate more than I'll eat. It's a pretty woodland plant, better considered a wildflower than a wild food. Menominee used *U. grandiflora* folklorically for swellings; Ojibwa for pleurisy and stomachache; Potawatomi for backache, myalgia, and tendonitis. Iroquois used *U. perfoliata* root tea for broken bones, cough, and sore eyes. Cherokee used *U. sessilifolia* for boils and diarrhea,

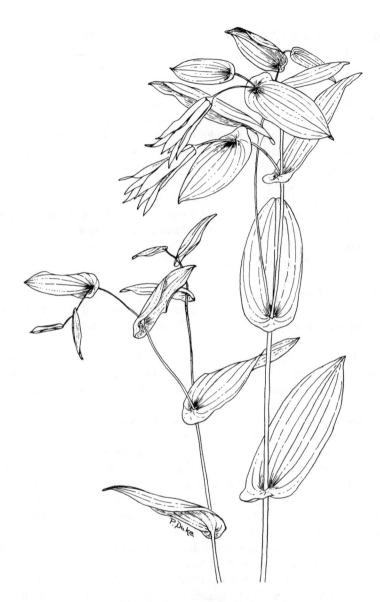

Uvularia perfoliata L.

Iroquois for blood disorders and broken bones (Moerman, 1986). WSSA lists only *U. sessilifolia* L., the <u>Little Bellwort</u>, as a weed, and that's the only species recommended as food by Facciola (1990).

VACCINIUM STAMINEUM L.   (ERICACEAE) ---
Gooseberry, Squaw Huckleberry, Wild Blueberry

DESCRIPTION: Deciduous shrub 2–3 feet tall. LEAVES: one at the node, alternating along the stem, narrowly egg-shaped, 1–4 inches long, 0.5–1.5 inches wide, pointed at the tip, almost toothless on the margin, but with a few glandular teeth near the base; broadest below, or rarely at or above the middle; rounded to the short leafstalk, often with a waxy coating that can be rubbed off with the fingers; midrib with several lateral veins heading to the margins. FLOWERS: white, seemingly solitary in the angle formed by smaller leaves (bracts) with the zigzag stem, drooping; sepals 5, basally united; petals 5, united below; stamens 10, sticking out of the flower (not the case in most blueberries); ovary 5-celled, fused with the sepals. FRUIT: a berry, turning greenish, pinkish, or purplish, tart.

DISTRIBUTION: Common shrub in dry thickets, in open deciduous forests or at the edges of coniferous forests. Flowering from April to June, with the fruits ripening a couple months later. KS-MA-OK-FL. Zones 4–8.

UTILITY: I've selected this species as one of the more common and easily recognized species of the wild blueberries, this one also called buckberry, deerberry, and gooseberry. All have edible fruits. All can be eaten as is, added to fruit cocktails, or made into jams, jellies, juices, or pies. Jellies of *Vaccinium stamineum* are said to have a unique amber green color (Facciola). Phenologists might note that in Maryland, July 4, 1991, most tame and wild blueberries were ripe or overripe, but not the gooseberry. Still green! Leaves of many blueberry species have been suggested, in teas, to correct hyperglycemia. Many other medicinal virtues, real or imagined, are ascribed to blueberry leaf tea. Recently I was called by two lawyers on opposite sides of a case in which an allegedly alcoholic Indian blamed an alleged murder on alleged insanity following ingestion of alleged blueberry root tea. They asked about chemicals in blueberry roots that might exacerbate the effects of alcoholism. I told both lawyers that I had no data on the chemistry of the roots. Then I referred to Erichsen-Brown (1979). Was I embarrassed! She had much more data! Citing Gunn (1861), speaking of blueberries in general, she learned that the root was the "part generally used in medicine". The astringent, diuretic root decoction was used for bowel disorders, including diarrhea, and was gargled for sore throat and mouth, and for indolent ulcers. And the clincher: "The roots and berries, bruised and tinctured in gin are a good diuretic, and seldom fail to relieve gravelly and dropsical affections; to be drank freely, or as much as the stomach and head will bear." Quoting Millspaugh (1892): "Said to be narcotic — a property also ascribed to the wine of Whortleberries (*V. uliginosum*) which is very intoxicating." Quoting

Vaccinium sp.

Densmore, she noted that Iroquois placed dried flowers of the low bush blueberry, *V. angustifolium,* on hot stones and inhaled the fumes for "craziness". Quoting Rafinesque, re the berries, "The Indians made a kind of wine of them. Indians smoked leaves of some blueberries as well . . . ." WSSA lists as weeds only *V. uliginosum* L. (Alpine Blueberry) and *V. vacillans* Torr. (Low Blueberry).

VIOLA PAPILIONACEA Pursh.   (VIOLACEAE) ---
Common Blue Violet, Common Violet, Meadow Violet

DESCRIPTION: Hairless, stemless perennial herbs, the leaf- and flower stalks all arising from ground level, from an elongate, overwintering rootstock. LEAVES: Heart-shaped, hairless green blades up to 5 inches long, 4 inches broad, the leafstalk even longer than the blade, most of the veins arising from the notch in the "heart", with rounded teeth on the margins. FLOWERS: Long-stalked, violet, pea-shaped flowers that are bilaterally symmetrical, with a 5-pointed green star between the flower stalk and the purple flower. FRUIT: A small, green, many-seeded, football-like pod with three sutures, the minute seeds olive to brown or black.

DISTRIBUTION: Damp alluvial forests, flower gardens, hedgerows, lawns, pastures, and roadsides, from full shade to nearly full sun, a nuisance weed mostly in perennial flower gardens. Occasionally flowering during warm spells in the winter farther south, the violets mostly start flowering in March slowly down by June in the Carolinas and Maryland. MN-MA-OK-GA. Zones 3–8.

UTILITY: I have eaten and enjoyed the flowers of this violet and the weedy johnny-jump-ups, but violets really are marginal food species. I have also enjoyed jellies and herbal vinegars made from violet flowers. As early as 1828–1830, writing in America, Rafinesque noted that violet flowers were used for a "grateful tea and syrup, used for cough, sorethroat, constipation, often given to children" (Erichsen-Brown, 1979). Violets are now a favorite among the flower eaters. But there are many flowers I prefer. Yanovsky mentions that *Viola pedunculata* leaves were used as greens by California Indians. Facciola states that the leaves are cooked and served like spinach. Moerman (1986) devotes more than two pages to violets, some of which are reported to contain up to 4,000 ppm salicylic acid, a compound also reported from willow bark. No wonder the Cherokee use violets for colds and headache. Erichsen-Brown reports 2,640 ppm vitamin C (ascorbic acid) and erroneously 200,000 ppm vitamin A. Zennie and Ogzewalla (1977) reported 15,000–20,000 IU vitamin A per 100 grams, which translates to 9,000–12,000 micrograms/100 grams or 90–120 ppm beta-carotene, still quite impressive. Having obtained carotene values for 10 wild edibles, they showed that violet, daisy, lambsquarter, garlic mustard, groundivy, plantain, and wild garlic contained more beta-carotene than spinach, richest of the widely marketed cultivated potherbs. Cultivated vegetables often have higher moisture percentages than their wilder relatives. Like buckwheat flowers, violet flowers can provide biologically significant quantities of rutin, which strengthens capillary blood vessels, CAUTION: Some of the foraging books mention the

Viola papilionacea Pursh.

rootstocks as edible. My few experiments have been equivocal, but I think the violet rootstock is mildly cathartic, and I cannot endorse it as food. I believe Rafinesque when he suggests that all our 40 species have rather similar properties, among them cathartic and emetic roots. WSSA also lists as weeds *V. arvensis* Murr. (Field Violet), *V. lanceolata* L. (Lanceleaf Violet), *V. odorata* (English Violet), and *V. rafinesquii* Greene (Field Pansy).

VITIS SPP. (VITACEAE) --- Wild Grapes

DESCRIPTION: High-climbing, deciduous, often woody vines, with tendrils. LEAVES: one at the node, alternating on the stem, 2–10 inches long, 1–8 inches broad, almost round in outline, but often deeply lobed; pointed at the tip; deeply and irregularly toothed at the margins, notched at the heart-shaped base, with 3–7 veins branching out at the base, and a few more laterals off the midrib farther up, with a tendril on the opposite side of the stems from some leaves. FLOWERS: small, numerous in tight clusters, the stalks of the cluster often on the opposite side of the stem from a leaf stalk; sepals 5, united; petals 5; stamens 5; ovary 2-celled, with 1 terminal process. FRUIT: a 1- to 4-seeded berry, reddish, purplish, black, or, in cultivated escapes, green.

DISTRIBUTION: Fencerows, old fields, moist thickets, deciduous and coniferous forests, especially alluvial forests, the woody stem divorced from the trees into which they have climbed (as opposed to poison ivy, which clings to the trees it climbs, with numerous aerial roots which penetrate the bark), flowering May to June, fruiting from August to frost, then sometimes fermenting on the vine. WA-ME-CA-FL. Zones 3–9.

UTILITY: Grapes provide both food and water. As far as I can learn, the only woody vine with stringy brown bark, hanging several feet away from the trunks of the trees it climbs, at least here in Maryland piedmont forest, is the grape vine. If you need water, you can fill your canteen with a vine 1–2 inches in diameter, making a first cut about 6 feet below where you will make your second cut. With your canteen ready, make the second cut, breaking the hundreds of minute straws with which the leaves are ''sucking'' water from the soil. Easy, but you have just wiped out a marvelous old grapevine, whose grapes, tiny or large, could have given you more beverage than that canteenful. Sample fruits of all the species; some are great, others mediocre, others nearly disgusting. The young, tender leaves and shoots are avidly eaten by visitors from the Middle East, whether the grape be wild or cultivated. The seeds, normally spit out, probably contain more nourishment, in the way of fats and proteins, than the fruit. Sun-dry your grapes to make your own raisins, being careful to avoid fungal infection. The distribution given above is rather generic, including both tame and wild grapes, all of which can be used. Generically, all seem to have edible fruits, and many seem to have palatable leaves. WSSA lists as weeds *Vitis aestivalis* Michx. (Summer Grape), *V. labrusca* L. (Fox Grape), *V. mustangensis* Buckl. (Mustang Grape), and *V. vulpina* L. (Frost Grape). CAUTION: If the woody vine hugs the tree,

Vitis sp.

with little rootlike structures attaching the vine to the tree trunk, be wary. It could even be poison ivy, and nobody wants to suck on a poison ivy vine. Watch for poison ivy clinging to your firewood, as well. Smoke from poison ivy can cause the allergy, too. Specimens more than 100 years old can, too.

YUCCA FILAMENTOSA L. (LILIACEAE) --- Adam's Needles; Spanish Bayonet

DESCRIPTION: Stemless or short-stemmed evergreen perennials. LEAVES: clustered at the tip of the stem or at ground level, like long, green swords, 12–32 inches long, 1–3.5 inches wide, pointed at the tip, toothless on the margins, but with curling fibers sticking out there, clasping at the bases, with numerous parallel veins, running almost the length of the leaf. FLOWERS: in elongate clusters on stalks overtowering the leaves, white or greenish; sepals 3; petals 3, broader than the sepals; stamens 6, shorter than the petals; ovary 3-celled. FRUIT: a many-seeded, oblong pod 0.5–2.0 inches long, the seeds black, flat, ca. 1/4 inch long.

DISTRIBUTION: Old fields, roadsides, railroad rights-of-way, apparently "happy" in sandy soils, flowering in April to June (Carolina to Maryland), the fruits ripening from September to October. PA-NJ-AL-FL. Zones 5–9.

UTILITY: I have eaten the flower buds raw, in May, while hiking along the railroad track. They have the strong taste of raw string beans to me. Fernald and Kinsey say no more than "It is frequently stated that fresh flowers of Yucca, properly dressed, are a good salad." In Latin America, I enjoyed the flowers of *Yucca elephantipes,* served in omelets and soups (Duke, 1972). Other species of *Yucca,* and perhaps local varieties of this one, have succulent fruits, the pulp of which, but not the seeds and fiber, can be cooked and eaten. Yanovsky says Indians ate fruits of this species. They have an even stronger raw-green bean flavor than the flowers. Recently, I read that flowers of Yucca contain significant quantities of estrogenlike compounds. While I noticed no change of voice or sexual desire following my feast of yucca buds, readers, especially those in need of estrogens, might be advised, that in Russia, flowers of this species are being considered as a possible starting point in the synthesis of steroids. Catawba Indians rubbed the roots onto dermatitis, perhaps taking advantage of the steroids. Cherokee took the root infusion for diabetes, and poulticed the beaten root as a salve onto sores. They also used the plant somehow as a sedative. Nanticoke poulticed the roots onto sprains (Moerman, 1986). WSSA lists as weeds only some western species: *Y. elata* Engelm. (Soaptree Yucca) and *Y. glauca* Nutt. ex Fraser (Great Plains Yucca). Both the saponins and the Blackfeet Indian usage of the plant as a tonic for falling hair may have contributed to the introduction of Yucca saponins into shampoos and the like. Cheyenne used it for dandruff, scabs, skin ailments, and to prevent baldness. CAUTION: Saponins and estrogens can be harmful in large quantities. *Y. glauca* was somehow used as a "delirifacient". Navajo held some species to be poisonous (Moerman, 1986).

Yucca filamentosa L.

A BIBLIOGRAPHY OF EDIBLE PLANTS
(Asterisked References* Cited in Text)

Angier, B. 1974. *Field Guide to Edible Wild Plants*. Stackpole Books, Harrisburg, PA.

Baird, E. A. and Lane, M. G. 1947. The seasonal variation in the ascorbic acid content of edible wild plants commonly found in New Brunswick. *Can. J. Res.*

Balls, E. K. 1965. Early uses of California plants. *California Natural History Guide 10*. University of California Press, Berkeley.

Baranov, A. I. 1967. Wild vegetables of the Chinese in Manchuria. *Econ. Bot.* 12:140–155.

Bean, L. J. and Saubel, K. S. 1972. Temalpakh. Cahuilla Indian knowledge and usage of plants. Malki Museum Press, Morongo Indian Reservation, CA.

Beardsley, G. 1939. The groundnut as used by the Indians of eastern North America. *Pap. Mich. Acad. Sci. Arts Lett.* 25:507–525.

Benson, E. M., Peters, J. M., Edwards, M. A., and Hogan, L. A. 1973. Nutritive values of wild edible plants of the Pacific northwest. *J. Am. Diet. Assoc.* 62:143–147.

Berkes, F. and Farkas, C. S. 1978. Eastern James Bay Cree Indians: changing patterns of wild food use and nutrition. *Ecol. Food Nutr.* 7:155–172.

*Brill, S. No date. *Shoots and Greens of Early Spring in Eastern North America.* "Wildman" Steve Brill, Publisher, Jamaica, NY. 55 pp.

Brown, D. K. 1954. Vitamin, protein, and carbohydrate content of some Arctic plants from the Fort Churchill, Manitoba, region. *Can. Def. Res. Board Pap.* 23:1–12.

Brown, R. 1868. On the vegetable products used by the Northwest American Indians, as food and medicine, in the arts, and in superstitious rites. *Trans. Bot. Soc. Edinburgh.* 9:378–396.

Canada Department of Health and Welfare. 1971. Indian Food: a Cookbook of Native Foods from British Columbia. Medical Services Branch, Pacific Region, Vancouver.

Casetter, E. F. 1935. Uncultivated native plants used as sources of food. New Mexico University Bulletin No. 266. (*Biol. Ser.* 4(1)).

CF = *Coltsfoot; In Appreciation of Wild Plants;* [Six bimonthly issues (Volume 12 in 1991). Subscriptions $10.00/year. c/o Jim Troy, Rt. 1, Box 313A, Shipman, VA 22971.]

*Clarke, C. B. 1977. *Edible and Useful Plants of California.* University of California Press, Berkeley. 280 pp.

*Croom, E. M. 1982. Medicinal Plants of the Lumbee Indians. Unpublished doctoral dissertation. North Carolina State University, Raleigh.

Core, E. L. 1967. Ethnobotany of the Southern Appalachian Aborigines. *Econ. Bot.* 21:199–214.

*CSIR (Council of Scientific and Industrial Research). 1948–1976. *The Wealth of India.* 11 volumes. New Delhi.

Densmore, F. 1928. Uses of Plants by the Chippewa Indians. U.S. Bur. Am. Ethnol. 44th Annu. Rep. 1926, 27:275–397.

*Dharmananda, S. 1988. Kudzu. *Bestways* (August 1988): 52–55.

Dore, W. G. 1970. A wild ground-bean, *Amphicarpa,* for the garden. *Greenhouse Garden Grass,* 9(2):7–11.

Douglas, F. H. 1931. Iroquois Foods. Department of Indian Art, Denver Art Museum, Leaflet No. 26.

Draper, H. H. 1977. The aboriginal Eskimo diet in modern perspective. *Am. Anthropol.* 79:309–316.

Drury, H. F. and Smith, S. G. 1956. Alaskan wild plants as an emergency food source. Science in Alaska. Proc. 4th Alaskan Science Conference, 1953. Juneau, Alaska. 155–159.

Duke, J. A. 1961. *Survival Manual I: South Vietnam.* Published by the author, Roneo, Durham, NC.

Duke, J. A. 1967. *Darien Survival Manual.* Battelle Memorial Institute, Columbus, OH. 80 pp.

Duke, J. A. 1970. Darienta's Dietary. Bioenvironmental and Radiological-Safety Feasibility Studies. Atlantic-Pacific Interoceanic Canal. Battelle Memorial Institute, Columbus, OH. U.S. AEC Rep. BMI-171–31. 111 pp.

*Duke, J. A. 1972. *Isthmian Ethnobotanical Dictionary.* Harrod and Co. Baltimore, MD. 96 pp.

Duke, J. A. 1975. Ethnobotanical observations on the Cuna Indians. *Econ. Bot.* 29(3):278–293.

*Duke, J. A. 1976. Quack salad and cancer, in *Natural Healing,* Bricklin, Mark, Ed., Rodale Press, Emmaus, PA. 249–251.

*Duke, J. A. 1976. Perennial weeds as indicators of annual climatic parameters. *Agric. Meteorol.* 16:291–294.

Duke, J. A. 1977. Vegetarian vitachart. *Q. J. Crude Drug Res.* 15:45–66.

Duke, J. A. 1983. The marvelous mayapple. *The Botanical Grower.* 1(1):3–4. [Newsletter]

Duke, J. A. 1981. *Handbook of Legumes of World Economic Importance.* Plenum Press, New York. 345 pp.

Duke, J. A. 1983. Incredible wild rice (*Zizania aquatica*) — native American food staple. *The Herb Report.* 3(2):2. [Newsletter]

Duke, J. A. 1984. Properties of the groundnut. In *The International Permaculture Seed Yearbook 1984,* Ed. D. Hemenway. 27–29.

Duke, J. A. 1984. *Apios* as an experimental "animal". *The International Permaculture Seed Yearbook 1984,* Ed. D. Hemenway. 29.

Duke, J. A. 1985. *CRC Handbook of Medicinal Herbs.* CRC Press, Boca Raton, FL. 677 pp.

Duke, J. A. 1985. An Herb a Day....The currant vs the evening primrose. *The Business of Herbs.* 3(3):12–13, July/August.

*Duke, J. A. 1986. *Handbook of Northeastern Indian Medicinal Plants.* Quarterman Publications, Lincoln, MA. 212 pp.

Duke, J. A. 1986. A crop to grow between the rows. *Curran's Ginseng Farmer.* 6(11):2–3.

Duke, J. A. 1986. Herbal water purification? HerbalGram 3(1):3,9. [Letter]

Duke, J. A. 1986. *Isthmian Ethnobotanical Dictionary,* 3rd ed. Scientific Publishers. Jodhpur, India. 205 pp.

Duke, J. A. 1986. Foraging items the Amerindians did not eat. *Coltsfoot.* 7(6):4–6.

Duke, J. A. 1987. An Herb a Day...Chicory. *The Business of Herbs.* 5(3):14, July/August.

Duke, J. A. 1987. An Herb a Day...Wintergreen. *The Business of Herbs.* 4(6):8–9, January/February.

Duke, J. A. 1988. Evening primrose: more American than apple pie — A Fourth-of-July retraction. *Coltsfoot.* 9(4):14–15, July/August.

Duke, J. A. 1988. An Herb a Day...Plantain. *The Business of Herbs.* 6(3):7–8, July/August.

Duke, J. A. 1989. On plantain. "Herbs" — Their growing and use. *Voice of the Texas Herb Growers,* 2(1):1,3.

Duke, J. A. 1989. An Herb a Day...Cranberry. *The Business of Herbs.* 7(5):6–7, November/December.

Duke, J. A. 1989. Orthomegalomania. *Organica* 8(28):12–13. Aubrey Organics, Tampa, FL.

Duke, J. A. 1989. Foods as pharmaceuticals, in *Herbs '89. Proceedings of the Fourth National Herb Growing and Marketing Conference:* IHGMA Conference, Simon, J. E., Kestner, A., and Buehrle, M. A., Eds., San Jose, CA, July 22–25. 166–176

*Duke, J. A. 1990. An Herb a Day...Red Clover. *The Business of Herbs.* 8(4):8–9. September/October.

Duke, J. A. 1990. An Herb a Day...Maypop (Passionflower). *The Business of Herbs.* 8(2):6–7. May/June.

*Duke, J. A. 1990. Take time to eat the roses. Bits of BARC [Beltsville USDA Newsletter] (November/December).

Duke, J. A. and Atchley, A. A. 1986. *CRC Handbook of Proximate Analysis Tables of Higher Plants.* CRC Press, Inc., Boca Raton, FL. 389 pp.

*Duke, J. A. and Ayensu, E. S. 1985. *Medicinal Plants of China.* 2 vols. Reference Publications, Algonac, MI. 705 pp.

*Duke, J. A. and Barnett, R. 1989. Cole's role. *Am. Health* 8(8):100, 102.

Duke, J. A. and Meares, P. 1983. Evening primrose. *The Business of Herbs.* July/August. 4–5. [Newsletter].

*Duke, J. A. and Wain, K. 1981. Medicinal Plants of the World. Computerized Index on Medicinal Plants and their Folk Medicinal Virtues. Computer Printout.

*Elias, T. S. and Dykeman, P. A. 1982. *Field Guide to North American Edible Wild Plants.* Van Nostrand Reinhold Co., New York. 286.

*Elliott, D. B. 1976. *Roots. An Underground Botany and Forager's Guide.* The Chatham Press, Old Greenwich, CT.

The Environmentarian. [Monthly Newsletter issued by Wildfood Lady, Linda Runyon (Phoenix, AZ), with columns by Jim Duke and Tom Squirer.]

*Erichsen-Brown, C. 1979. *Use of Plants for the Past 500 Years.* Breezy Creeks Press, Aurora, Canada. 512 pp.

*Facciola, S. 1990. *Cornucopia—A Source Book of Edible Plants.* Kampong Publications, Vista, CA. 678 pp.

Fenton, W. N., Ed. 1968. *Parker on the Iroquois,* Syracuse University Press, Syracuse, NY.

*Fernald, M. L. and Kinsey, A. C. 1958. *Edible Wild Plants of Eastern North America.* Revised by Reed C. Rollins. Harper and Row, New York.

*Foster, S. 1984. *Herbal Bounty.* Peregrine Smith Books, Salt Lake City. 192 pp.

*Foster, S. and Duke, J. A. 1990. *Peterson Field Guide to Medicinal Plants.* Houghton Mifflin, Boston. 366 pp.

French, D. H. 1965. Ethnobotany of the Pacific Northwest Indians. *Econ. Bot.* 19:378–382.

Gaertner, E. E. 1967. *Harvest without Planting.* E.E. Gaertner, Chalk River, Ontario.

Gibbons, E. 1962. *Stalking the Wild Asparagus.* David McKay Co., New York.

Gibbons, E. 1966. *Stalking the Healthful Herbs.* David McKay Co., New York.

Gibbons, E. 1973. *Stalking the West's Wild Foods.* National Geographic. 144:187–199.

Gillespie, W. H. 1959. *A Compilation of the Edible Wild Plants of West Virginia.* Press of Scholar's Library, New York. 118 pp.

Gilmore, M. R. 1919. Uses of plants by the Indians of the Missouri River region. U.S. Bur. Am. Ethnol., 33rd Annu. Rep. 1911, 12:53–154.

*Grieve, M. 1974. *A Modern Herbal* (Reprint). Hafner Press, New York. 915 pp.

Hall, A. 1973. *The Wild Food Trailguide.* Holt, Rinehart, and Winston, New York. 195 pp.

Hamel, P. B. and Chiltoskey, U. 1975. *Cherokee Plants and Their Uses: a 400 Year History.* Herald Publishing Co., Sylva, NC.

*Harris, B. C. 1971. *Eat the Weeds.* Barre Publishers, Barre, MA. 223 pp.

Harris, G. H. 1890 (1891). Root foods of the Seneca Indians. *Proc. Rochester Acad. Sci.* 1:106–115.

*Hartwell, Jonathan L. 1984. *Plants Used Against Cancer.* Quarterman Publications, Lincoln, MA.

Havard, V. 1895. Food plants of the North American Indians. *Bull. Torrey Bot. Club.* 22:98–123.

Havard, V. 1896. Drink plants of the North American Indians. *Bull. Torrey Bot. Club.* 23:33–46.

*Hays, S. M. 1991. Ten weeds we could live without. Agricultural Research USDA, June 1991. 4–9.

*Heinerman, J. 1988. *Heinerman's Encyclopedia of Fruits, Vegetables and Herbs.* Parker Publishing Co., West Nyack, NY. 400 pp.

Heller, C. A. 1966. *Wild Edible and Poisonous Plants of Alaska.* Cooperative Extension Service, University of Alaska, Publ. No. 28.

Hellson, J. C. and Gadd, M. 1974. Ethnobotany of the Blackfoot Indians. Can. Ethnol. Serv. Pap. No. 19, Natl. Mus. Man, Mercury Series. National Museums of Canada, Ottawa.

Hitchcock, S. T. 1980. *Gather Ye Wild Things.* Harper and Row, New York. 180 pp.

*Hobbs, C. and Foster, S. 1990. Hawthorn: A Literature Review. Herbalgram #22: 18–33.

*Holm, L. G., Plucknett, D. L., Pancho, J. V., and Herberger, J. P. 1977. *The World's Worst Weeds*. University Press of Hawaii, Honolulu. 609 pp.

Hussey, J. S. 1974. Some useful plants of early New England. *Econ. Bot.* 28:311–337.

*Jones, P. 1991. *Just Weeds, History, Myths and Uses*. Prentice-Hall, New York. 303 pp.

Keays, J. L. 1976. Foliage. Part I. Practical utilization of foliage. *Appl. Polym. Symp.* 28:445–464.

*Kindscher, K. 1987. *Edible Wild Plants of the Prairie. An Ethnobotanical Guide*. University Press of Kansas, Lawrence. 276 pp.

*Kingsbury, J. M. 1964. *Poisonous Plants of the United States and Canada*. Prentice-Hall, Englewood Cliffs, NJ.

Krochmal, C. and Krochmal, A. 1974. *A Naturalist's Guide to Cooking with Wild Plants*. Quadrangle, New York.

Kuhnlein, H. V. and Calloway, D. H. 1977. Contemporary Hopi food intake patterns. *Ecol. Food Nutr.* 6:159–173.

Laferriere, J. E. 1989. Nutricomp Program [A Data Base of Nutritional Compositions of Hundreds of Foods]. Dept. Ecol. and Evolutionary Biol., University of Arizona, Tucson, 85721. Reviewed in *J. Ethnobiol.* 9(1): 27–29. 1989.

*Leung, A. Y. 1980. *Encyclopedia of Common Natural Ingredients*. John Wiley & Sons, New York. 409 pp.

*Leung, A.Y. 1984. *Chinese Herbal Remedies*. Universe Books, New York. 192 pp.

*Martin, A. C., Zim, H. S., and Nelson, A. L. 1951. *American Wildlife & Plants*. Reprint: 1961. Dover Publications, New York. 500 pp.

Martin, F. W. and Ruberte, R. M. 1975. *Edible Leaves of the Tropics*. Antillian College Press, Mayaguez, PR.

*Meares, P. 1981. The ageless elder. *Country* (July 1981). 48–51

*Mitchell, J. C. and Rook, A. 1979. *Botanical Dermatology*. Greenglass, Vancouver. 787 pp.

*Mitich, L. Series of articles over several years in *Weed Technology*.

*Moerman, D. E. 1986. *Medicinal Plants of Native America*. 2 volumes. University of Michigan Museum of Anthropology.Tech. Rep. No. 19. Ann Arbor

Morton, J. F. 1963. *Wild Plants for Survival in South Florida*. Hurricane Press, Miami.

Morton, J. F. 1963. Principal wild food plants of the United States. *Econ. Bot.* 17:319–330.

*Morton, J. F. 1981. *Atlas of Medicinal Plants of Middle America*. Charles C Thomas, Springfield, IL. 1,420 pp.

Moulton, L. 1979. *Herb Walk 1*. The Gluten Co., Provo, UT. 398 pp.

*Naegele, T. A., 1980. *Edible and Medicinal Plants of the Great Lakes*. Survival Seminars, Calumet, MI. 427 pp.

Newberry, J. S. 1887. Food and fiber plants of North American Indians. *Pop. Sci. Mon.* 32:32–46.

Nickerson, N. H., Rose, N. H., and Richter, E. A. 1973. Native plants in the diets of North Alaskan Eskimos, in *Man and His Foods*. Smith, C. E., Jr., Ed. University of Alabama Press, University, AL. pp. 3–27.

Norman, H. 1991. Personal communication.

Palmer, E. 1871. Food products of the North American Indians. U.S. Dep. Agric. Rep. 1870:404–428.

Palmer, E. 1878. Plants used by the Indians of the United States. *Am. Nat.* 12:593–606, 646–655.

Palmer, G. 1975. Shuswap Indian ethnobotany. *Syesis*, 8:29–81.

Peterson, L. 1978. *A Field Guide to Edible Wild Plants of Eastern and Central North America*. Houghton Mifflin, Boston. 330 pp.

Porsild, A. E. 1937. Edible Roots and Berries of Northern Canada. Natl. Mus. Can. Dep. Mines Res., Ottawa.

*Rafinesque, C. S. 1828–1830. *Medical Flora or Manual of Medical Botany of the United States*. Vol. 1. Atkinson & Alexander, Philadelphia. Vol. 2. S. C. Atkinson, Philadelphia.

Robbins, W. W., Harrington, J. P., and Freire-Marreco, B. 1916. Ethnobotany of the Tewa Indians. U.S. Government Printing Office, Washington, D.C.

Robson, J. R. K. and Elias, J. N. 1978. *The Nutritional Value of Indigenous Wild Plants; an Annotated Bibliography.* Whitston Publishing Co., Troy, N.Y.

*Rogers, B. and Powers-Rogers, B. 1988. *Culinary Botany.* PRP-Powers, Rogers & Plants, Kent, WA. 176 pp.

Rudin, D. and Felix, C. 1987. The Omega-3 Phenomenon.

Runyon, L. *The Environmentarian.* [Monthly Newsletter on Edible Plants. Wild Foods Inc., 3531 W. Glendale Ave., Suite 369, Phoenix, AZ 85051.]

Saunders, C. F. 1920. *Useful Wild Plants of the United States and Canada.* Robert M. McBride, New York.

Smith, H. H. 1923. *Ethnobotany of the Menomini Indians.* Bull. Publ. Mus., City of Milwaukee, No. 1:1–174.

Smith, H. H. 1928. *Ethnobotany of the Meskwaki Indians.* Bull. Publ. Mus., City of Milwaukee, No. 4:175–326.

Smith, H. H. 1932. *Ethnobotany of the Ojibwe Indians.* Bull. Publ. Mus., City of Milwaukee, No. 4:327–525.

Smith, H. H. 1933. *Ethnobotany of the Forest Potawatomi Indians.* Bull. Publ. Mus., City of Milwaukee, No. 7:1–230.

Speck, F. G. and Dexter, R. W. 1951. Utilization of animals and plants by the Micmac Indians of New Brunswick. *J. Wash. Acad. Sci.* 41:250–261.

Squier, T. 1991. *Wild Foods Are Good Foods.* WFF (Wild Foods Forum) 2(1): 7–8

Stevenson, M. C. 1915. Ethnobotany of the Zuni Indians. U.S. Bur. Am. Ethnol., 30th Annu. Rep. 1908, 09:33–102.

*Stitt, P. A. 1990. *Why George Should Eat His Broccoli.* Dougherty Co., Milwaukee. 399 pp.

Stout, A. B. 1914. Vegetable foods of the American Indians. *J. N.Y. Bot. Gard.* 15:50–60.

Stubbs, R. D. 1966. An Investigation of the Edible and Medicinal Plants Used by the Flathead Indians. M.A. thesis, University of Montana, Missoula.

Sturtevant, E. L. 1919. Sturtevant's Notes on Edible Plants. Edited by U.P. Hedrick, New York Department of Agriculture, Annu. Rep. 27, Vol. 2(2). [Reprint 1972 as *Sturtevant's Edible Plants of the World.* Dover Publications, New York.]

Szczawinski, A. F. and Turner, N. J. 1978. Edible garden weeds of Canada. Edible Wild Plants of Canada, No. 1. Natl. Mus. Nat. Sci., Natl. Mus. Can., Ottawa.

Szczawinski, A. F. and Turner, N. J. 1980. Wild green vegetables of Canada. Edible Wild Plants of Canada, No. 4. Natl. Mus. Nat. Sci., Natl. Mus. Can., Ottawa.

*Tanaka, T. 1976. *Tanaka's Cyclopedia of Edible Plants of the World.* Keigaku Publishing, Tokyo. 924 pp.

Tomikel, J. 1973. *Edible Wild Plants of Pennsylvania and New York.* Allegheny Press, Pittsburgh. 88 pp.

Turner, N. J. 1973. The ethnobotany of the Bella Coola Indians of British Columbia. *Syesis,* 6:193–220.

Turner, N. J. 1975. Food Plants of British Columbia Indians. Part 1. Coastal Peoples. B.C. Prov. Mus. Handb. No. 34, Victoria.

Turner, N. J. 1978. Food Plants of British Columbia Indians. Part 2. Interior Peoples. B.C. Prov. Mus. Handb. No. 36, Victoria.

Turner, N. J. and Bell, M. A. M. 1971. The ethnobotany of the Coast Salish Indians of Vancouver Island. *Econ. Bot.* 25:63–104.

Turner, N. J. and Bell, M. A. M. 1971. The ethnobotany of the southern Kwakiutl Indians of British Columbia. *Econ. Bot.* 27:257–310.

Turner, N. J., Bouchard, R., and Kennedy, D. I. D. 1980. Ethnobotany of the Okanagan-Colville Indians of British Columbia and Washington. B.C. Prov. Mus. Occas. Pap. No. 21, Victoria.

Turner, N. J. 1981. A gift for the taking: the untapped potential of some food plants of North American native peoples. *Can. J. Bot.* 59(11):2331–2357.

Turner, N. J. and Efrat, B. S. 1982. Ethnobotany of the Hesquiat people of the west coast of Vancouver Island, B.C. Prov. Mus. Cultural Recovery Pap. No. 2, Victoria.

Turner, N. J. and Szczawinski, A. F. 1979. Edible wild fruits and nuts of Canada. Edible wild plants of Canada, No. 3. Natl. Mus. Nat. Sci., Natl. Mus. Can., Ottawa.

*Uphof, J. C. Th. 1968. *Dictionary of Economic Plants*. Verlag von J. Cramer. Lehre. 591 pp.

Virginia Polytechnic Institute. 1968. Edible fruits of forest trees. Publ. 184, Extension Division, V.P.I., Blacksburg, VA.

*Watt, J. M. and Breyer-Brandwijk, M. G. 1962. *The Medicinal and Poisonous Plants of Southern and Eastern Africa*. E. & S. Livingstone, Ltd., Edinburgh. 1457 pp.

Waugh, F. W. 1916. Iroquois foods and food preparation. Geol. Sur. Can. Mem. 86 (Anthropol. Ser. 12). (Facsimile edition 1973.)

WT = *WEED TECHNOLOGY* (WT). WSSA

WFF = *WILD FOODS FORUM*. [Monthly newsletter put out by Deborah Duchon, 4 Carlisle Way, N.E., Atlanta, GA 30308. Subscription $15/year.]

Williams, K. 1977. *Eating Wild Plants*. Mountain Press Publishing, Missoula, MT. 180 pp.

Wilson, T. 1916. The use of wild plants as food by Indians. *Ottawa Nat.* 30(2):17–21.

Wittrock, M. A. and Wittrock, G. L. 1942. Food plants of the Indians. *J. N.Y. Bot. Gard.* 43:57–71.

*Yanovsky, E. 1936. Food Plants of the North American Indians. U.S. Department of Agriculture, Misc. Publ. 237, 1–88.

Yanovsky, E. and Kingsbury, R. M. 1938. Analyses of some Indian food plants. *J. Assoc. Off. Agric. Chem.* 21:648–665.

*Zennie, T. M. and Ogzewalla, C. D. 1977. Ascorbic acid and vitamin A content of edible wild plants of Ohio and Kentucky. *Econ. Bot.* 31(1):76–79.

LIST OF ILLUSTRATIONS

Indexes

INDEX TO PLANT NAMES

In the text, 100 genera of weeds are arranged, alphabetically by generic names. Most of you readers, not familiar with generic names, will find the common names in the Index to Plant Names. I have used the scientific generic name rather than the highly variable and sometimes confusing array of common names. In most cases, more than one edible species occurs in the United States representatives of the genus, though the account may single out what I view to be most important. For some of these genera, one can find dozens of common names, but I have chosen what I believe to be the most important or well-studied species, using one or more of the more prevalent common names. That common name may belong to all species in the genus or just one species in the genus.

GENERAL INDEX

A

Abortion, 194
Abscesses, 76, 82, 145, 154, 178
Acetogenins, 42
Acid soils, 172
Acne, 158
Acorns, 164
Acupuncture, 171, 205
Addiction, 164
Ague, 65
AIDS, 149, 159
Alabama Indians, 128
Alcoholic beverages, 19, 26, 48, 92, 160,
    208–209, see also Beer; Wine
Ale, 70, 204
Alexiterics, 72, 143–144
Algonquin, 140
Alkaline soils, 92
Alkaloids, 38–39, 67, 110–111, 117, 146,
    164, 201
Allantoin, 151
Allelochemicals, 116
Allergies, 27, 70, 167, 213
Allopurinol, 142
Alluvial areas, 24, 42, 48, 92, 110, 114,
    152, 154, 176, 204, 206, 210, 212
Alpha-linolenic-acid (ALA), 6, 10, 142
Alpha-tocopherol, 156
Alteratives, 82, 94, 101
Alternative fuels, 8, 144
Ameba, 178
Analgesics, 82, 122, 144
Anesthetics, 38
Angelica candy, 80
Anodyne, 66, 94, 110, 143
Anthrax, 157
Antidotes, 132, 141, 143, 172, 180–181
Antiemetic, 140
Antihistamines, 82
Antilactigogue, 181
Antioxidants, 30
Antiscorbutics, 58
Antiseptic, 18, 26, 132, 142, 143, 154
Antispasmodic, 94
Antitussive, 143

Anxiety, 159
Apache, 108, 144
Aperient, 72, 101
Apertif, 46, 101, 120, 122
Aphrodisiac, 76, 80, 120, 140
Appendicitis, 154
Appetite stimulator, 58
Aquatics, 18, 22, 34, 38, 58, 130, 132–
    137, 144, 174, 178, 202
Araucano, 52
Arid lands, 116
Artemisinin, 4–5, 9
Arthritis, 63, 102–105, 126, 145, 148,
    154, 204
Asarone, 18–19
Ascites, 22, 76
Ascorbic acid, 50, see also Vitamin C
Ashes, 56–57, 156
Asthma, 70, 121, 131, 135, 143, 200
Astringent, 20, 60, 70, 82, 94, 96, 98,
    122, 181, 208
Atherosclerosis, 26
Atropine, 106, 147
Awobana paper, 76

B

Baby powder, 65
Backache, 135, 158, 206
Bactericides, 82, 126
Bacteriostatic, 132
Bait, 18
Baking soda, 36
Baldness, 131, 178, 214
Balds, 30
Balsams, 46
Barnyards, 54, 118, 142, 158, 194
Basketry, 38
Beach plants, 66
Bear, 104
Beaver, 49, 60, 136, 140, 174
Beer, 20, 66, 86, 88, 106, 112, 188, see
    also Root beer
Bee stings, 46, 150
Bella Coola, 118, 158
Benzaldehyde, 31
Beri-beri, 22

T

U

V

W